Edited by Roman Grigoriev

**Transport and Mixing in
Laminar Flows**

Related Titles

Gilmore, R., Lefranc, M.

The Topology of Chaos

Alice in Stretch and Squeezeland

2012

ISBN: 978-3-527-41067-5

Colin, S. (ed.)

Microfluidics

2010

ISBN: 978-1-84821-097-4

Zikanov, O.

Essential Computational Fluid Dynamics

2010

ISBN: 978-0-470-42329-5

Fabry, P., Fouletier, J.

Chemical and Biological Microsensors

Applications in Fluid Media

2009

ISBN: 978-1-84821-142-1

Ansorge, R., Sonar, T.

Mathematical Models of Fluid Dynamics

Modeling, Theory, Basic Numerical Facts - An Introduction

2009

ISBN: 978-3-527-40774-3

Hsu, T.-R.

MEMS & Microsystems

Design, Manufacture, and Nanoscale Engineering

2008

ISBN: 978-0-470-08301-7

Löhner, R.

Applied Computational Fluid Dynamics Techniques

An Introduction Based on Finite Element Methods

2007

ISBN: 978-0-470-51907-3

Gomez, F. A. (ed.)

Biological Applications of Microfluidics

2008

ISBN: 978-0-470-07483-1

Das, S. K., Choi, S. U., Yu, W., Pradeep, T.

Nanofluids

Science and Technology

2008

ISBN: 978-0-470-07473-2

Munson, B. R., Young, D. F., Okiishi, T. H.

Fundamentals of Fluid Mechanics

2005

ISBN: 978-0-471-67582-2

Edited by Roman Grigoriev

Transport and Mixing in Laminar Flows

From Microfluidics to Oceanic Currents

WILEY-VCH Verlag GmbH & Co. KGaA

The Editors

Prof. Roman Grigoriev
Georgia Inst. of Technology
School of Physics
837 State Street
Atlanta, GA 30332-0430
USA

All books published by **Wiley-VCH** are carefully produced. Nevertheless, authors, editors, and publisher do not warrant the information contained in these books, including this book, to be free of errors. Readers are advised to keep in mind that statements, data, illustrations, procedural details or other items may inadvertently be inaccurate.

Library of Congress Card No.: applied for

British Library Cataloguing-in-Publication Data
A catalogue record for this book is available from the British Library.

Bibliographic information published by the Deutsche Nationalbibliothek
The Deutsche Nationalbibliothek lists this publication in the Deutsche Nationalbibliografie; detailed bibliographic data are available on the Internet at http://dnb.d-nb.de.

© 2012 Wiley-VCH Verlag & Co. KGaA, Boschstr. 12, 69469 Weinheim, Germany

All rights reserved (including those of translation into other languages). No part of this book may be reproduced in any form – by photoprinting, microfilm, or any other means – nor transmitted or translated into a machine language without written permission from the publishers. Registered names, trademarks, etc. used in this book, even when not specifically marked as such, are not to be considered unprotected by law.

Typesetting   Thomson Digital, Noida, India
Printing and Binding   Fabulous Printers Pte Ltd, Singapore
Cover Design   Adam-Design, Weinheim

Printed in Singapore
Printed on acid-free paper

Print ISBN: 978-3-527-41011-8
ePDF ISBN: 978-3-527-63976-2
oBook ISBN: 978-3-527-63974-8
ePub ISBN: 978-3-527-63975-5
Mobi ISBN: 978-3-527-63977-9

Contents

List of Contributors IX

Mixing in Laminar Fluid Flows: From Microfluidics to Oceanic Currents 1
Roman O. Grigoriev
Introduction 1

1	**Resonances and Mixing in Near-Integrable Volume-Preserving Systems** 5
	Dmitri Vainchtein
1.1	Introduction 5
1.2	General Properties of Near-Integrable Flows and Different Types of the Resonance Surfaces 8
1.2.1	Metrics of Mixing 10
1.2.2	Correlations of Successive Jumps and Ergodicity 12
1.3	Separatrix Crossings in Volume-Preserving Systems 12
1.3.1	Flow Structure 14
1.3.2	Dynamics Near the Separatrix Surface 15
1.3.3	Finite Perturbations 16
1.4	Passages Through Resonances in Autonomous Flows 17
1.4.1	Scattering on Resonance 19
1.4.2	Capture Into Resonance 20
1.4.3	Improved AI 21
1.4.4	Jump of AI Between First- and Second-Layer Boundaries 22
1.4.5	Long-Time Dynamics and Adiabatic Diffusion 23
1.5	Passages Through Resonances in Nonautonomous Flows 24
1.5.1	Unperturbed Flow 25
1.5.2	Two Perturbations and Averaging 26
1.5.3	Resonant Phenomena 27
1.5.4	Volume of the Mixing Domain 29
	References 31

2	**Fluid Stirring in a Tilted Rotating Tank** *35*
	Thomas Ward
2.1	Introduction and Background Information *35*
2.2	Tilted-Rotating Tank Analysis *38*
2.2.1	Tilted-Rotating Tank Model Equation *38*
2.2.1.1	Asymptotic Analysis: Free Surface Vortex *40*
2.2.1.2	Linear Analysis: Periodic Shearing *42*
2.2.2	Comments on Laminar Flow *44*
2.2.3	Analytical Results *45*
2.3	Experiments *47*
2.3.1	Setup and Procedure *47*
2.3.1.1	Low Reynolds Number Experiments: Homogeneous Fluid *47*
2.3.1.2	Laminar Flow Experiments: Homogeneous and Inhomogeneous Fluids *48*
2.3.2	Results and Analysis *49*
2.3.2.1	Low Reynolds Number *49*
2.3.2.2	Laminar Flow: Homogeneous Fluid *51*
2.3.2.3	Laminar Flow: Inhomogeneous Fluid *53*
2.3.3	Brief Discussion *55*
2.4	Conclusion *55*
	References *56*
3	**Lagrangian Coherent Structures** *59*
	Shawn C. Shadden
3.1	Introduction *59*
3.2	Background *61*
3.3	Global Approach *64*
3.3.1	FTLE *65*
3.3.2	FTLE Ridges *67*
3.3.3	Nature of Stretching *68*
3.3.4	Objectivity *71*
3.4	Computational Strategy *72*
3.4.1	Grid-Based Computation *74*
3.4.2	Integration Time *75*
3.4.3	LCS Extraction *79*
3.5	Robustness *82*
3.6	Applications *83*
3.7	Conclusions *83*
	References *84*
4	**Interfacial Transfer from Stirred Laminar Flows** *91*
	Joseph D. Kirtland and Abraham D. Stroock
4.1	Introduction *91*
4.2	Phenomena and Definitions *92*
4.3	Experimental Methods *95*

4.3.1	Protein Binding	95
4.3.2	Electrochemical Reactions	95
4.3.3	Heat Transfer from Macroscopic Coiled Pipe	97
4.3.4	Interphase Mass Transfer from Droplets	98
4.3.5	Summary of Experimental Observations	99
4.4	Modeling Approaches	100
4.4.1	Numerical Solutions in Eulerian Frame	100
4.4.2	Numerical Solutions in Lagrangian Frame	101
4.4.3	Macrotransport Approach	103
4.4.4	Theoretical Approaches	103
4.5	Conclusions	107
	References	107
5	**The Effects of Laminar Mixing on Reaction Fronts and Patterns**	**111**
	Tom Solomon	
5.1	Introduction	111
5.2	Background	113
5.2.1	Laminar mixing – the Advection–Diffusion Equation	113
5.2.2	Short-Range Mixing	113
5.2.3	Long-Range Transport of Impurities	115
5.2.4	Nonlinear Reactions	117
5.2.5	Reaction–Diffusion Systems	117
5.3	Advection–Reaction–Diffusion: General Principles	118
5.4	Local Behavior of ARD Systems	120
5.5	Synchronization of Oscillating Reactions	122
5.6	Front Propagation in ARD Systems	125
5.7	Additional Comments	127
	References	128
6	**Microfluidic Flows of Viscoelastic Fluids**	**131**
	Mónica S. N. Oliveira, Manuel A. Alves, and Fernando T. Pinho	
6.1	Introduction	131
6.1.1	Objectives and Organization of the Chapter	131
6.1.2	Microfluidics	131
6.1.2.1	Basic Principles, Relevance, and Applications	131
6.1.2.2	Complex Fluids in Microfluidic Flows	135
6.1.2.3	Continuum Approximation	136
6.2	Mixing in Microfluidics	138
6.2.1	Challenges of Micromixing	138
6.2.2	Overview of Methods for Micromixing Enhancement	139
6.3	Non-Newtonian Viscoelastic Fluids	141
6.3.1	Shear Viscosity	142
6.3.2	Normal Stresses	143
6.3.3	Storage and Loss Moduli	144
6.3.4	Extensional Viscosity	144

6.3.5	Other Rheological Properties	145
6.4	Governing Equations	146
6.4.1	Continuity and Momentum Equations	146
6.4.2	Rheological Constitutive Equation	147
6.4.2.1	Generalized Newtonian Fluid Model	148
6.4.2.2	Viscoelastic Stress Models	148
6.4.3	Equations for Electro-Osmosis	151
6.4.4	Thermal Energy Equation	153
6.5	Passive Mixing for Viscoelastic Fluids: Purely Elastic Flow Instabilities	154
6.5.1	General Considerations	154
6.5.2	The Underlying Physics	155
6.5.3	Viscoelastic Instabilities in Some Canonical Flows	156
6.5.4	Elastic Turbulence	160
6.6	Other Forcing Methods - Applications	162
6.6.1	Electro-Osmosis	162
6.6.2	Electrophoresis	164
6.7	Conclusions and Perspectives	165
	References	166

Index *175*

List of Contributors

Manuel A. Alves
Universidade do Porto
Faculdade de Engenharia
Centro de Estudos de Fenómenos
de Transporte
Departamento de Engenharia Química
Rua Dr. Roberto Frias s/n
4200-465 Porto
Portugal

Roman O. Grigoriev
Georgia Institute of Technology
School of Physics
837 State St.
Atlanta, GA 30332-0430
USA

Joseph D. Kirtland
Dordt College
Department of Physics
Sioux Center, IA 51250
USA

Mónica S. N. Oliveira
Universidade do Porto
Faculdade de Engenharia
Centro de Estudos de Fenómenos
de Transporte
Departamento de Engenharia Química
Rua Dr. Roberto Frias s/n
4200-465 Porto
Portugal

Fernando T. Pinho
Universidade do Porto
Faculdade de Engenharia
Centro de Estudos de Fenómenos
de Transporte
Departamento de Engenharia Mecânica
Rua Dr. Roberto Frias s/n
4200-465 Porto
Portugal

Shawn C. Shadden
Illinois Institute of Technology
Mechanical, Materials & Aerospace
Engineering
10 West 32nd Street
243 Engineering 1 Building
Chicago, IL 60616-3793
USA

Tom Solomon
Bucknell University
Department of Physics & Astronomy
Lewisburg, PA 17837
USA

Abraham D. Stroock
Cornell University
School of Chemical and Biomolecular
Engineering
120 Olin Hall
Ithaca, NY 14853
USA

Thomas Ward
North Carolina State University
Department of Mechanical and
Aerospace Engineering
911 Oval Dr.
Raleigh, NC 27695-7910
USA

Dmitri Vainchtein
Temple University
Department of Mechanical Engineering
1947 N 12th St.
Philadelphia, PA 19122
USA

Mixing in Laminar Fluid Flows: From Microfluidics to Oceanic Currents

Roman O. Grigoriev

Introduction

Transport properties of laminar fluid flows have attracted a lot of attention over the past decade. To a large extent, this increased interest has been driven by the rapid development of microfluidic technologies which revolutionized practical molecular biology by enabling high-throughput sequencing. Speeding up the scaled-down versions of chemical essays, such as polymerase chain reaction, demanded by applications in genomics, proteomics, and clinical pathology requires quick and thorough mixing of several components (e.g., DNA fragments and fluorescent markers) inside microscopic liquid volumes. Although molecular diffusion becomes an effective mixing mechanism at small scales, it is not fast enough at the scale of a typical microfluidic device. Advection by a moving fluid provides an alternative, and much quicker, mixing mechanism. This conclusion follows directly from the advection–diffusion equation

$$\partial_t c = -\mathbf{v} \cdot \nabla c + D\nabla^2 c \tag{1}$$

which describes the evolution of solute concentration $c(\mathbf{x}, t)$ with molecular diffusion constant D in the fluid solvent moving with velocity $\mathbf{v}(\mathbf{x}, t)$. Nondimensionalizing this equation we discover that, for a fluid flow with the characteristic length scale L and velocity scale V, molecular diffusion described by the $D\nabla^2 c$ term is dominant when the nondimensional Peclet number $Pe \equiv VL/D$ is small, while advection described by the $\mathbf{v} \cdot \nabla c$ term dominates in the opposite limit. Typical microfluidic devices happen to be characterized by $Pe < 1$, placing them in the diffusion-dominated regime.

We should also point out that the evolution equation (1), where the concentration c is replaced by the temperature T of the fluid and the molecular diffusivity D is replaced by the heat diffusivity κ, describes heat transport in the fluid, with numerous modern applications in microelectronics cooling. At the opposite end of the size scale, the same equation has been used to describe the transport properties of atmospheric and oceanic currents. Some of the high-profile applications include problems describing the spreading of environmental pollutants such as oil and radioactive elements released as a result of accidents at offshore drilling platforms and nuclear plants.

A mathematically equivalent, but more analytically tractable, description of advective transport in the limit of negligible diffusion is provided by replacing partial differential equation (1) with an ordinary differential equation

$$\dot{\mathbf{x}}(t) = \mathbf{v}(\mathbf{x}, t) \tag{2}$$

which defines the streamlines of the flow, but also describes the motion of infinitesimal fluid elements characterized by a constant value of concentration c (or temperature T) on time scales on which diffusion can be neglected. The Jacobian $\nabla \mathbf{v}$ of the system (2) describes the local shape dynamics of the fluid elements in the local comoving reference frame, as they are stretched in some directions and compressed in others, as explained in Chapter 3. This process is usually referred to as *Lagrangian transport*, reflecting the choice of the Lagrangian reference frame comoving with the flow, as opposed to the static global Eulerian reference frame used in (1).

While this reduction in the transport problem affords tremendous simplification, the system (2) defines a set of three coupled nonlinear ordinary differential equations, which in general can only be solved numerically. One exception is the special limit of a steady integrable flow $\mathbf{v}(\mathbf{x})$ which possesses one or two conserved quantities, referred to as integrals of motion, $\Phi(\mathbf{x}(t)) = $ const. These integrals arise when the flow possesses a symmetry. At the microscale, symmetries are pervasive and usually reflect the geometry of the flow (e.g., spherical shape of microdroplets, round or rectangular shape of microchannels, etc.). The flows with one invariant, also known as *action–angle–angle* flows, are effectively two-dimensional (with quasi-periodic streamlines), while the flows with two invariants, of the *action–action–angle* type are one-dimensional (with time-periodic streamlines). Integrable flows possess extremely poor mixing properties because level sets of each invariant represent a transport barrier – a two-dimensional surface which cannot be crossed by any streamline of the flow.

Both types of integrable flows allow exact analytical solutions to be found, although rarely in explicit form. Solutions to nonintegrable flows are chaotic and cannot be found in analytic form. However, approximate semianalytical short-term solutions can be computed, for both steady and time-periodic flows, using perturbation theory described in Chapter 1 when the velocity field is near-integrable, that is, $\mathbf{v}(\mathbf{x}, t) = \mathbf{v}_0(\mathbf{x}) + \mathbf{v}_1(\mathbf{x}, t)$, where $\mathbf{v}_0$ is an integrable flow and $|\mathbf{v}_1| \ll |\mathbf{v}_0|$. The perturbative analysis shows that, somewhat surprisingly, the simpler action–action–angle flows possess dramatically better mixing properties than the more complex action–angle–angle flows in the presence of a generic perturbation. This is illustrated in Chapter 2 which compares theoretical predictions with experimental observations for a perturbed action–action–angle flow arising in a tilted rotated tank.

While the perturbative description of advective transport is unique in that it allows one to describe the mixing process quantitatively, it breaks down in the nonperturbative regime. There are two leading alternatives which have been used to characterize mixing in the general case; both providing a qualitative description. One, analytical, is based on the topological analysis of the streamlines of the flow in the

Eulerian reference frame. Another, numerical, is based on finite-time Lyapunov exponents which measure the rate of stretching or contraction of infinitesimal fluid elements in the Lagrangian reference frame. As Chapter 3 explains, the *Lagrangian coherent structures* defining regions of the flow characterized by the largest or the smallest values of Lyapunov exponents tend to organize the transport. Specifically, the largest exponents define the locations of the transport barriers and the smallest – the regions of the flow where mixing is most efficient.

Advective transport discussed in Chapters 1–3 describes only the initial stage of mixing. For a flow with chaotic streamlines, the fluid elements are stretched in some directions and, at the same time, compressed in others exponentially fast. The associated sharpening of the concentration gradients leads to a greatly enhanced molecular diffusion (or heat conduction). Therefore, on longer time scales, one inevitably has to revert to using the advection–diffusion equation (1). Moreover, diffusive transport might be non-negligible even at earlier times, either for sufficiently small volumes or in the regions of the flow where the velocity of the fluid becomes very small (e.g., near solid walls). The latter situation is discussed in Chapter 4, which considers transport across interfaces, either between fluids (characterized by stress-free boundary conditions) or between a fluid and a solid (no-slip boundary conditions). Understanding transport across interfaces is essential in problems involving, for example, heat exchange between fluids and solids or chemical reactions localized to the walls.

Chapter 5 is devoted to the interplay between mixing and chemical reactions taking place in the bulk. Indeed, chemical reaction rates depend on the local concentrations of reactants and products, and hence transport phenomena are expected to play an important role. The most general models of chemical reactors are formulated in terms of *reaction–advection–diffusion* equations which generalize (1):

$$\partial_t c_i = F_i - \mathbf{v} \cdot \nabla c_i + D_i \nabla^2 c_i \tag{3}$$

where the nonlinear functions $F_i(c_1, \ldots, c_k)$ describe the reaction rates, $c_1, \ldots, c_k$ are the concentrations of different chemicals, and D_i – the corresponding diffusion constants. While the field of reaction–diffusion systems is well established – the Turing mechanism for pattern formation has been proposed in 1952 – no comprehensive theory of reaction–advection–diffusion system currently exists. Experiments described in Chapter 5 show that the fluid flow, especially when it is time-dependent, has a highly nontrivial effect on the dynamics of reaction fronts, illustrating the importance of both transport mechanisms – diffusion and advection – on the chemical reaction patterns arising in fluids. In particular, transport barriers are determined to play a crucial role in the organization of chemical patterns.

Quite generally, and especially at the microscale, steady flows are found to have inferior mixing properties, compared with their time-dependent counterparts. This is quite logical since mixing is caused by stretching and compression of fluid elements advected along chaotic streamlines. According to dynamical systems theory, in order for chaos to arise, the flow has to be at least three-dimensional. On the other hand, both symmetries and spatial confinement tend to reduce the effective dimensionality of the flow, making it two- or even one-dimensional and hence precluding chaotic streamlines. The effective dimensionality of the flow can be increased either by removing spatial confinement and symmetry (forcing the fluid through a twisted microchannel or a microchannel with a variable cross-section are good examples) or by introducing time-dependence.

Microscale flows of *Newtonian fluids* such as water, however, tend to be steady. Indeed, the flow of Newtonian fluids is governed by the Navier–Stokes equation

$$\varrho(\partial_t \mathbf{v} + \mathbf{v} \cdot \nabla \mathbf{v}) = -\nabla p + \mu \nabla^2 \mathbf{v} + \mathbf{f} \tag{4}$$

where ϱ is the density of the fluid, μ is its dynamic viscosity, p is the pressure, and $\mathbf{f}$ is the body force density. At the microscale, the Reynolds number $\mathrm{Re} = \varrho V L/\mu$ is usually much smaller than unity, so (4) reduces to the Stokes equation

$$\mu \nabla \times \nabla^2 \mathbf{v} = -\nabla \times \mathbf{f} \tag{5}$$

Hence, the direction of the driving force $\mathbf{f}$ has to be time-dependent in order to generate an unsteady flow $\mathbf{v}$, which can be quite hard to achieve in practice.

An alternative approach, described in Chapter 6, is to use *non-Newtonian* – or viscoelastic – fluids, which are characterized by both viscous and elastic stresses. Elastic stresses are represented by additional terms on the right-hand-side of (4) which can remain non-negligible even at the typical length scales of microfluidic devices. As a result, in certain geometries the flow becomes susceptible to viscoelastic instabilities which introduce time-dependence, dramatically enhancing mixing. Understanding transport in non-Newtonian fluids is interesting not only from the fundamental perspective, but also due to numerous emerging applications such as clinical pathology which involve handling of polymeric fluids, for example, whole blood or protein solutions.

While this book does not pretend to provide complete coverage of the subject of transport and mixing in laminar fluid flows, it provides a review of the main research thrusts in this active field. It is our hope that the book will appeal to both scientists working on the fundamental aspects of the transport problem and the engineers looking to use the existing knowledge base in the design of a new generation of microfluidic devices.

1
Resonances and Mixing in Near-Integrable Volume-Preserving Systems
Dmitri Vainchtein

1.1
Introduction

Many laminar flows are often characterized by a high degree of symmetry due to the confining effect of surface tension (for free-surface flows, e.g., in microdroplets) and/or device geometry (e.g., for flows in microchannels). Designing a flow with good mixing properties is particularly difficult in the presence of symmetries. Symmetry leads to the existence of (flow) *invariants* [1, 2], which are functions of coordinates that are constant along streamlines of the flow. The level sets of one invariant define surfaces on which the (three-dimensional) flow is effectively two-dimensional. An additional invariant further reduces the flow dimensionality: a flow with two invariants is effectively one-dimensional. Since the flow cannot cross invariant surfaces, the existence of invariants is highly undesirable in the mixing problem as their presence inhibits complete stirring of the full fluid volume by advection. Neither is chaotic advection per se sufficient for good mixing, as time-dependent flows [3, 4] can have chaotic streamlines restricted to two-dimensional surfaces in the presence of an invariant. Thus, the key to achieving effective chaotic mixing in any laminar flow is to ensure that all flow invariants are destroyed.

In this section we will focus on the class of laminar flows characterized by small deviations from exact symmetries. Not only are such flows common in various applications of microfluidics, this is *the only* class of flows that generically affords a quantitative analytical treatment. The description of the weakly perturbed flow in terms of the action and angle variables allows quantitative analytical treatment using perturbation theory. Indeed, if the symmetries are broken weakly, the invariants (or actions) of the unperturbed flow become slowly varying functions of time (start to *drift*, in the more technical language) for the perturbed flow, while the angle variable remains quickly varying. Such perturbed flows are referred to as *near-integrable*, in contrast to the flows with exact symmetries which are *integrable*, that is, possess an exact analytical solution. Near-integrable systems play a prominent role in many areas of science. Often they arise naturally when there is a large separation of scales and, hence, of the associated forces, for example, as in many problems in celestial mechanics, where the gravitational interaction with the Sun dominates all other forces which can be considered small perturbations [5, 6]. Similarly, for weakly perturbed action–action–angle fluid flows there is a large separation of timescales on which the actions and the angle change.

The space of the integrable unperturbed system is foliated into invariant tori and the motion on these tori is quasi-periodic or periodic. If there are two independent integrals, the tori are invariant closed curves. In general, the integrability requires the existence of at least one *conserved quantity* (or *action* or *invariant*), so all flows of interest belong to one of two classes: action–action–angle or action–angle–angle [1]. Transport in the perturbed action–angle–angle flows is severely restricted by KAM tori (it was illustrated in [7]), while the effective degeneracy of the action–action–angle flows opens the possibility of global transport and mixing. We will, therefore, focus our attention on action–action–angle flows and possible mechanisms leading to chaotic advection.

Exact analytic solutions for near-integrable dynamics cannot be obtained. Direct brute-force numerical simulation of such systems is possible, but usually very challenging precisely due to a big separation of timescales. Approximate analytical tools represent an important alternative for studying such systems. Specifically, the assumption of a weak perturbation allows one to use a collection of perturbation theory methods to describe the dynamics quantitatively. In particular, by averaging the evolution equations for the actions I and J over a period of the fast motion described by the angle ϕ one finds that although the original exact invariants are destroyed, the averaged system of equations itself possesses an invariant $\Phi(I, J)$. Since the averaged equations are an approximation, in the exact perturbed system Φ is only conserved in the adiabatic sense: its value undergoes small oscillations with period close to that of ϕ but the average value of Φ remains the same on much longer time intervals [8]. Therefore, this approximate invariant is referred to as an *adiabatic invariant* (AI).

As it turns out, the existence of an AI enables a greatly simplified description of the mixing dynamics in near-integrable flows. Of course, if the AI were conserved everywhere, mixing would be restricted to the two-dimensional level sets of the AI (usually tori) defined by $\Phi = \text{const}$, which is often indeed the case (i.e., nonintegrability does not lead to mixing). Mixing requires the breakdown of the adiabatic invariance, which can occur in the flows possessing certain types of *singular manifolds*, where the fast subsystem slows down and the separation of scales disappears. A very elegant description of the dynamics can be obtained by separating the evolution into regular advection along the 2D level sets of the AI *between* singular manifolds, and fast passages *through* the singular manifolds. During the motion near the singular manifolds, the value of the AI typically experiences a change that is much larger than the magnitude of the oscillations of the AI during the motion far from the singular manifolds. As the time of the passage near the singular manifolds is much shorter than the characteristic time of motion between them, the changes in the AI can be treated as instantaneous *jumps* in describing the evolution of the AI. For every set of initial conditions, the magnitude of the jump in the AI can be calculated exactly.

However, a small change in the initial conditions produces in general a large change in the jump magnitude [9, 10]. Hence, for weak perturbations, in computing the statistical properties of many consequent jumps, it is possible to treat the jump magnitude as a random variable with statistical properties obtained from the dependence of the jump magnitude on initial conditions. The dynamics in the vicinity of singular manifolds (separatrices or resonance surfaces) can be described using a different perturbation expansion, where the small parameters are not just the perturbation strength but also the distance to the singular manifold. For resonant surfaces, this approach was first introduced in [11, 12] and further developed in [10, 13] in the context of Hamiltonian systems and subsequently applied to 3D volume-preserving autonomous (such as flows of incompressible fluids) in [14] and nonautonomous [4] systems. A theory for systems with separatrix crossings was proposed in [15, 16] and later developed in [17–19] for Hamiltonian systems and in [9, 20] for 3D volume-preserving autonomous systems.

If allowed by the geometry of the system, the streamline keeps coming to the singular surface(s) again and again and the process of jumps repeats itself. Accumulation of jumps at multiple crossings results in destruction of the adiabatic invariance (i.e., the AI changes by a value of the order 1) and leads to chaotic dynamics in the system. Therefore, the physical space becomes partitioned into the domains of chaotic and regular dynamics filled, respectively, by the streamlines that do, or do not, cross the singular manifold(s). In the chaotic domain, the jumps of the AI associated with separatrix or resonance crossings lead to the destruction of adiabatic invariance and transport across the level sets of the AI. Over long timescales, accumulation of small jumps coupled with the divergence of initially close streamlines lead to effective diffusion of the AI and mixing. This dynamical picture allows one to compute the size and shape of the chaotic and regular domains and to estimate the rate of mixing.

The theory of long-time transport in volume-preserving flows in the presence of chaotic advection and regular diffusion is by no means limited to mixing in fluid flows [21–25]. Such a description of long-time transport in near-integrable Hamiltonian and volume-preserving systems is crucial for long-term predictions of, for example, the dynamics of comets and asteroids [26–28], customizing transport to achieve selective segregation in electromagnetic diverters in plasma confinement devices [29–31], energy exchange between coupled oscillators [32–34], chaotic billiards [35], arrays of Josephson junctions [36, 37], and the drift of charged particles in the Earth magnetosphere [38–42].

In this chapter, we describe destruction of AIs at separatrices and resonances and use several examples studied earlier [4, 14, 43, 44] to illustrate different aspects of the complete picture. We refer the reader to the corresponding chapter for details of derivations and additional discussions. General properties are discussed in Section 1.2. Separatrix crossings are discussed in Section 1.3, and passages through resonances in autonomous and nonautonomous flows are considered in Sections 1.4 and 1.5, respectively.

1.2
General Properties of Near-Integrable Flows and Different Types of the Resonance Surfaces

The motion of passive tracers advected by the flow can be described by a volume-preserving system of ODE in $\mathbf{R}^3$ depending on a small parameter, $0 < \varepsilon \ll 1$:

$$\dot{\mathbf{x}} = \mathbf{v}(\mathbf{x}) + \varepsilon \mathbf{w}(\mathbf{x}, t, \varepsilon), \qquad \text{div } \mathbf{v} = \text{div } \mathbf{w} = 0 \tag{1.1}$$

Velocity field $\mathbf{v}$ in (1.1) defines an unperturbed (base) flow; $\mathbf{w}$ is a perturbation and is supposed to be a smooth function of ε. We restrict our discussion to 3D autonomous base flows, while the perturbation may be autonomous or nonautonomous. System (1.1) at $\varepsilon = 0$ corresponds to the unperturbed system. In a sense, passive tracers in the flows are equivalent to phase points in generic dynamical systems.

The effects of the small perturbation in (1.1) start manifesting themselves on time intervals of order at least ε^{-1}. A function of phase variables is called an AI if its value along a phase trajectory of (1.1) has only small (with ε) variations on time intervals of such length. In other words, an AI is an approximate first integral of the system. Perpetual conservation of AI presents a barrier for complete mixing.

Let unperturbed system (1.1) be integrable and of the action–action–angle type. Then, almost the entire phase space is filled with closed streamlines. Denote the two independent integrals of motion as I and J. Every joint level of the two integrals $I = i$, $J = j$ defines a closed unperturbed phase trajectory $\Gamma_{i,j}$. Introduce on $\Gamma_{i,j}$ an angular variable ϕ mod 2π changing at a constant rate in the unperturbed motion.

The perturbation in (1.1) causes the values of I and J to change at a rate of order ε in the motion along a perturbed streamline. In terms of the variables i, j, ϕ, perturbed system (1.1) can be written as

$$\frac{di}{dt} = \varepsilon f(i, j, \phi, \varepsilon) \quad \frac{dj}{dt} = \varepsilon g(i, j, \phi, \varepsilon) \quad \frac{d\phi}{dt} = \Omega(i, j) + \varepsilon h(i, j, \phi, \varepsilon) \tag{1.2}$$

The functions f, g, h are 2π-periodic in ϕ. In (1.2), the variables i, j are "slow," and the variable ϕ is "fast." Define the averaged system:

$$\frac{di}{dt} = \varepsilon F(i, j) \quad \frac{dj}{dt} = \varepsilon G(i, j) \tag{1.3}$$

where functions F and G are obtained by averaging f and g, respectively, over ϕ:

$$F(i,j) = \frac{1}{2\pi} \int_{\Gamma_{i,j}} (\text{grad } I, \mathbf{w}) \, d\phi \quad G(i,j) = \frac{1}{2\pi} \int_{\Gamma_{i,j}} (\text{grad } J, \mathbf{w}) \, d\phi \tag{1.4}$$

In (1.4), $\mathbf{w}$ is calculated at $\varepsilon = 0$, parentheses denote the scalar product, and the integration is performed along $\Gamma_{i,j}$. Far from the singular surfaces (described below), solutions of the averaged system describe variations of i, j in complete system (1.1) with the accuracy of order ε on time intervals of order ε^{-1} [45, 46].

Let $\Phi(i, j)$ be the flux of the perturbation through a surface spanning $\Gamma_{i,j}$. Due to the preservation of the volume, the value of $\Phi(i, j)$ does not depend on a particular choice of the surface. A remarkable fact is that averaged system (1.3) is Hamiltonian, and

1.2 General Properties of Near-Integrable Flows and Different Types of the Resonance Surfaces

$\Phi(i,j)$ is a Hamiltonian function (see, e.g., [9, 20]):

$$\frac{di}{dt} = \frac{\varepsilon}{\mu(i,j)T(i,j)} \frac{\partial \Phi(i,j)}{\partial j}, \quad \frac{dj}{dt} = -\frac{\varepsilon}{\mu(i,j)T(i,j)} \frac{\partial \Phi(i,j)}{\partial i} \quad (1.5)$$

where $\mu(i,j)$ is a certain function of i and j, determined by the base flow and $T(i,j) = 2\pi/\Omega(i,j)$ is the period of the unperturbed motion along $\Gamma_{i,j}$. In almost all the systems we studied recently, $\mu(i,j) \equiv 1$. Moreover, it is always the case when the base flow has axial symmetry and the two invariants are the streamfunction and the azimuthal angle. It follows from (1.5) that $\Phi(i,j)$ is an integral of the averaged system. Standard assertions about the accuracy of the averaging method (see, e.g., [8, 45, 46]) imply that Φ is an approximate integral of the motion in exact system (1.1), that is, Φ is an *adiabatic invariant* (see Figure 1.1a below).

However, the averaging method breaks down in a neighborhood of 2D *singular surfaces*. These surfaces can be of one of three types:

1) *Separatrix surfaces* containing nondegenerated hyperbolic fixed points of the unperturbed system and filled by heteroclinic trajectories connecting them. A systems of this type is considered in Section 1.3.
2) *Separatrix surfaces* containing a line of degenerate singular points (this case occurs, in particular, in 1 d.o.f. Hamiltonian systems depending on a slowly varying parameter or a no-autonomous flows with axial symmetry, and it is considered elsewhere [15, 16, 47]).
3) *Resonance surfaces*, corresponding to closed curves filling a surface. In *autonomous* systems with one angle variable, the function $\Omega(i,j)$ in (1.2) is identically zero everywhere on such a surface. A system of this type is considered in Section 1.4. In *nonautonomous* systems in which time appears as an additional fast phase with a frequency ω, on resonance surface $\Omega(i,j)/\omega$ is a rational number (see Section 1.5). The major difference between separatrix and resonance surfaces is that near separatrix surfaces the base flow slows down only in

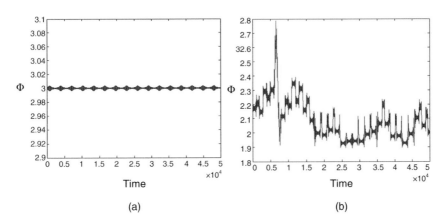

Figure 1.1 Adiabatic invariants (a) away from the singular surfaces and (b) when a streamline crosses them. The model from [14].

the immediate vicinity of fixed points of the base system. In comparison, a linear combination (with integer coefficients) of the phase of the base flow and the time variable slows down everywhere near a resonance surface.

Let us just briefly note that the problem of jumps of AIs at separatrix crossings in volume-preserving systems cannot be reduced to similar problems in Hamiltonian systems depending on a slowly varying parameter [15, 16, 47], or in slow–fast Hamiltonian systems [48]. Although ideologically close to them, this problem needs an independent study. Similar phenomena were also observed in 3D volume-preserving maps [7].

A complete description of chaotic advection in these problems starts with a description of a *single crossing* of a resonance or a separatrix surface. Let a phase point (passive tracer) $i(t), j(t)$ closely follow a trajectory of the averaged system. The quantity $\Phi(i,j)$ along the streamline oscillates with an amplitude of order ε around a certain constant value, say, Φ_1. When the streamline crosses a small neighborhood of a singular surface, Φ changes by a value $\Delta\Phi \sim \varepsilon^\alpha$, $0 < \alpha < 1$, which is in general much greater than ε. In the case of scatterings on resonance and almost all the separatrix crossings, $\alpha = 1/2$. After this neighborhood is crossed, the value of Φ along the trajectory oscillates near a new constant value, $\Phi_2 = \Phi_1 + \Delta\Phi$. As the main change occurs in a narrow neighborhood of a singular surface, we shall call this change a *jump of the AI*. In every particular problem, an asymptotic formula (in the limit of small values of ε) for this change of the AI can be obtained following a standard procedure which was reported in several publications [4, 9, 20, 43, 44, 49]. An example of such a dynamics is illustrated in (see Figure 1.1b). In the case of crossing a resonance when there is the possibility of capture into resonance, the captured dynamics can be also described.

The magnitude of a jump turns out to be very sensitive to variations of initial conditions. Therefore, the jump is in a sense random. If allowed by the geometry of the system, the streamline comes to the separatrix again and the process repeats itself. Accumulation of jumps at multiple crossings results in destruction of the adiabatic invariance (i.e., the AI changes by a value of the order 1) and leads to chaotic dynamics in the system. Based on the equations for a single passage, we can describe statistical properties of jumps and use them to study the long-time dynamics on time intervals that include many crossings.

1.2.1
Metrics of Mixing

Two different (and generally unrelated) metrics should be used to describe chaotic advection in a bounded flow such as the one considered here: the size of the chaotic domain and the characteristic rate of mixing inside the chaotic domain.

The first of these two metrics is *the volume of the chaotic domain*, V_c. We define the volume and the dimensionality of mixing as the properties of the domain occupied (after a long time) by the tracers that originate in a small ball (say, size ε). For $\varepsilon = 0$, the whole interior is almost completely regular for any kind of perturbation. Actually, any however minuscule perturbation (e.g., molecular diffusion) leads to 1D mixing: the dependence of $\Omega(i,j)$ in (1.2) on the values of I and/or J results in stretching of the original ball along $\Gamma_{i,j}$. This happens over order $O(1)$ times. In perturbed systems without singular surfaces, the tracers stay forever in the vicinity of the original surface of constant AI (see Figure 1.1a), eventually covering the whole surface $\Phi = \text{const}$. Thus we say that the adiabatic invariancy leads to 2D mixing.

In the presence of singular surfaces, a 3D chaotic domain of a finite size appears as soon as ε becomes nonzero. This is due to the fact that while the resonance phenomena are themselves local events (they are only affected by the dynamics in the vicinity of a corresponding surface), their effect is global, extending the chaotic domain to the scale of the entire flow. As a result, V_c depends on parameters of the base flow but not on the magnitude of perturbation, ε (for infinitesimal ε). Depending on parameters of the base flow (e.g., q in Section 1.3), the flow domain can be completely regular, partially regular and partially chaotic, or completely chaotic. The size of the chaotic domain is, to leading order, determined by the shape of its boundary – the torus τ_b tangential to the singular surface – which is independent of ε. However, for any finite ε, the boundary between the two domains is more complex. The chaotic domain penetrates inside τ_b adding a small layer (most often with thickness of order $\sqrt{\varepsilon}$, see [50] and Section 1.4 below for details). Further, small islands of stability may appear inside the mixing domain. As a result there are small corrections to V_c.

The second metric, the rate of mixing D, on the other hand, strongly depends on ε. Assuming statistical independence of consecutive crossings (see below), we can describe the evolution of the AI by a random walk with a characteristic step size of order ε^α, $0 < \alpha < 1$. Hence, after N crossings, the value of the AI changes by a quantity of the order $\sqrt{N} \times \varepsilon^\alpha$. The mixing can be considered complete when a typical chaotic streamline samples the entire chaotic domain. The difference between the values of the AI that bound the chaotic domain in our problem is of order unity. Therefore, it takes on the order of $N \sim \varepsilon^{-2\alpha}$ separatrix crossings for diffusion to cover the whole domain. As the typical time between successive crossings is of the order $1/\varepsilon$, we find the characteristic time for mixing to be $T_M = O(\varepsilon^{-1-2\alpha})$. This characteristic time diverges for $\varepsilon \to 0$, so the rate of mixing, defined as $D = 1/T_M = O(\varepsilon^{1+2\alpha})$, vanishes for $\varepsilon \to 0$.

For infinitesimal ε, the consecutive jumps can be considered statistically independent for most of the streamlines, so the accumulation of jumps can be described as a random walk *without* memory, leading to the standard Fokker–Planck equation for the probability density function (PDF) of the AI

$$\partial_t P = -\partial_\Phi (U_A P) + \partial_\Phi (D_A \partial_\Phi P) \tag{1.6}$$

where $P(\Phi, t)d\Phi$ is defined as the probability that at time t the tracer resides between the surfaces Φ and $\Phi + d\Phi$. The drift velocity U_A and the diffusion coefficient D_A are determined by the first and the second moments of the distribution of the jumps, respectively [51]. Since the jump magnitude distribution is a function of Φ, so are its moments. Moreover, since the dynamics between the jumps takes place on the surfaces of constant AI, the time between the jumps is also a function of Φ [43]. Therefore, both $U_A(\Phi)$ and $D_A(\Phi)$ should be computed by taking this dependence into account.

1.2.2
Correlations of Successive Jumps and Ergodicity

Quantitative properties of the diffusion of the AI, in particular the validity of (1.6), depend on whether consecutive crossings are statistically dependent or independent. In volume-preserving systems, statistical independence (and, thus, the hypothesis of ergodicity, at least up to a residual of a small measure) for vanishing value of ε can be deduced from the divergence of resonance phases (denoted by ξ in the following sections) along streamlines. A similar problem for the Hamiltonian system was discussed in [10, 52, 53]. For finite values of ε the consecutive jumps become somewhat correlated, especially near the boundaries of the system. We discuss this in more details in Section 1.3.

Consider statistical properties of the jumps in Φ along one-phase trajectory of the system. Let two successive separatrix crossings be characterized by values ξ_1 and ξ_2. A small variation $\delta \xi_1$ in ξ_1 produces a variation of the jump in Φ by $\sim \varepsilon^a \delta \xi_1$. In the period of time $\sim \varepsilon^{-1}$ before the next crossing, the value ξ_2 obtains a variation $\delta \xi_2 \sim \varepsilon^{a-1} \delta \xi_1$. Thus $\delta \xi_2 / \delta \xi_1 \sim \varepsilon^{a-1} \gg 1$. Therefore, it is natural to suppose that ξ_1 and ξ_2 are statistically independent and the successive jumps in Φ are not correlated.

For many flows, it was verified numerically that inside the chaotic domain one does indeed find a positive Lyapunov exponent for $\varepsilon > 0$, confirming the divergence of nearby streamlines. Furthermore, for small ε, the flow possesses good ergodic properties inside the mixing domain, as the Poincaré sections illustrate, indicating very thorough mixing. Indeed, the chaotic domain is essentially devoid of regular islands, so a single streamline densely fills the whole chaotic domain. For decreasing ε, the regular islands (of size $\sqrt{\varepsilon}$) are expected to gradually disappear, resulting in perfect mixing.

1.3
Separatrix Crossings in Volume-Preserving Systems

In this section, we consider a flow where the presence of separatrix crossings results in the destruction of adiabatic invariance to illustrate different aspects of the evolution. The following problem was studied in details in [43].

Consider a microdroplet suspended at the free surface of a liquid substrate and driven using the thermocapillary effect with a constant speed in a straight line along the substrate surface [3, 54]. Experiments found the mixing to be very poor in this regime [54]. However, a numerical study of the simplified model of the flow constructed in [3] shows that the mixing efficiency can be improved dramatically by appropriately choosing the parameters such as the magnitude of the temperature coefficients of surface tension at different fluid interfaces, the ratio $\lambda = \mu_{\text{in}} / \mu_{\text{out}}$ of the fluid viscosities inside and outside the droplet, and the curvature of the temperature field driving the flow.

To simplify the mathematical description of the problem, we follow [3, 43] assuming that the droplet is suspended below the free surface of the liquid substrate and consider the limit of small capillary numbers such that the droplet can be considered spherical. Under these assumptions, in nondimensional units (with distances scaled by the droplet radius and the origin located at the center of the drop), there are three component flows: the dipole flow $\mathbf{v}_d$

$$\begin{aligned}\dot{x}_d &= 1 + x^2 - 2r^2 \\ \dot{y}_d &= xy \\ \dot{z}_d &= xz\end{aligned} \tag{1.7}$$

the quadrupole flow $\mathbf{v}_q$

$$\begin{aligned}\dot{x}_q &= 2x(1 + x^2 - 2r^2) \\ \dot{y}_q &= y(r^2 + 2x^2 - 1) \\ \dot{z}_q &= z(r^2 + 2x^2 - 1)\end{aligned} \tag{1.8}$$

and the Taylor flow $\mathbf{v}_t$

$$\begin{aligned}\dot{x}_t &= z(\beta(5r^2 - 3 - 4x^2) + 2) \\ \dot{y}_t &= -4\beta xyz \\ \dot{z}_t &= x(\beta(5r^2 - 3 - 4z^2) - 2)\end{aligned} \tag{1.9}$$

where $r^2 = x^2 + y^2 + z^2$, $\beta = 1/(1+\lambda)$ (the value $\beta = 0.5$ is used in all numerical calculations), the x axis points in the direction of the thermal gradient, and the z axis is vertical. The components $\mathbf{v}_d$ and $\mathbf{v}_q$ are caused by the thermocapillary effect at the droplet surface, while $\mathbf{v}_t$ arises due to the thermocapillary effect at the surface of the liquid substrate. The complete flow inside the droplet can be written as a linear superposition of the dipole, Taylor, and quadrupole flows

$$\mathbf{v} = \mathbf{v}_d + \varepsilon \mathbf{v}_t + q\mathbf{v}_q \tag{1.10}$$

The parameters ε and q determine the relative strengths of the three components which depend on the temperature coefficients of surface tension at the droplet surface and the free surface of the substrate fluid and on the nonuniformity of the imposed temperature gradient [3]. As the dipole component is present in almost any setting, it is convenient to set its magnitude to unity by an appropriate choice of the timescale. In what follows, we will restrict our attention to the case of $|q| = O(1)$ and $0 \le \varepsilon \ll 1$. This will allow us to describe the mixing process quantitatively using perturbation theory [8].

Flows (1.7)–(1.10) are volume preserving and bounded by the droplet surface $r = 1$, which represents an invariant set. Moreover, the plane $y = 0$ is an invariant set for each flow. Since the flow for $y < 0$ is a mirror image of the flow for $y > 0$ and there is no transport across the $y = 0$ plane, we will restrict our attention to the flow inside the hemisphere characterized by positive values of y.

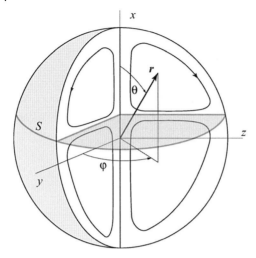

Figure 1.2 A sketch of the unperturbed flow for $|q| > 0.5$.

For $\varepsilon = 0$, flow (1.10) reduces to a superposition of the dipole and quadrupole flows and possesses two invariants: the azimuthal angle φ (around the x axis) and the streamfunction ψ:

$$\varphi = \arctan(z/y), \quad \psi = \frac{1}{2}(1 + 2qx)\varrho^2(1 - r^2)$$

where we have defined $\varrho^2 = y^2 + z^2$. The flow structure of the unperturbed system depends on the value of q. Note that the dynamics for $q > 0$ is the same as for $q < 0$ up to the reflection with respect to the plane $x = 0$. There is always a pair of hyperbolic fixed points at the poles $x = \pm 1$, $\varrho = 0$ and, for $|q| > 0.5$, two circles of degenerate elliptic fixed points, accompanied by a hyperbolic fixed point $x = x_s \equiv -1/(2q)$, $\varrho = 0$ and a circle of degenerate hyperbolic fixed points on the surface at $x = x_s$, $r = 1$. The plane $x = x_s$ (denoted S below) is a separatrix (see Figure 1.2). Away from the separatrix, the axis, and the surface of the sphere, each joint level of the two integrals φ and ψ defines a closed unperturbed phase trajectory $\Gamma_{\varphi,\psi}$. The motion on $\Gamma_{\varphi,\psi}$ is periodic with frequency $\Omega(\psi)$. Note that if the base flow possesses axial symmetry, its frequency is naturally independent on azimuthal angle.

1.3.1
Flow Structure

For $\varepsilon \neq 0$, system (1.10) is no longer integrable. Integrals ψ and φ are not preserved. Streamlines are not closed and cross the separatrix S. Figure 1.3a represents a result of long integration of one perturbed-phase trajectory.

The structure of the phase portrait on the slow (φ, ψ) plane depends on the values of q. For $|q| < 0.5$, there is no separatrix, so the averaging procedure is valid everywhere and hence the AI is constant, if one ignores small bounded oscillations

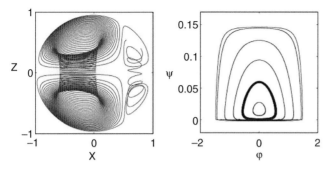

Figure 1.3 Dynamics of the perturbed system over one long period. The parameters are $q = -0.7$ and $= 10^{-2}$. (a) A perturbed streamline. (b) Phase portrait of the averaged system. The thick line in (b) shows the boundary of chaotic domain, l_b.

with amplitude of order ε. Therefore, the entire drop is a regular domain: all streamlines reside on the tori that are levels sets of the AI.

For $|q| > 0.5$, the separatrix plane defined by $x = x_s = -1/(2q)$ in the physical space (or $\psi = 0$ in the slow plane) appears inside the drop. For $0.5 < |q| < q_b$, the interior of the drop is divided between the regular domain and the chaotic domain. Numerically, we find $q_b \approx 0.96$. The regular domain corresponds to streamlines lying on the level sets of Φ that do not cross the separatrix (see Figure 1.3b), whereas the rest of the streamlines belong to the chaotic domain. To the leading order in ε, the boundary between the regular and the chaotic domain is a torus τ_b tangential to S. On the (φ, ψ) plane, τ_b corresponds to a closed curve l_b passing through the origin (see Figure 1.3b). Thus, we conclude that it is the level set $\Phi = \Phi(0,0)$ of the AI that serves as the boundary between the regular and the chaotic domains. Note that the level set $\Phi = \Phi(0,0)$ is not a sharp boundary: for any finite ε there are (although very few) regular trajectories inside the chaotic domain and vice versa. We will return to the discussion of the boundary between the domains in Section 1.4.

1.3.2
Dynamics Near the Separatrix Surface

For streamlines that cross the separatrix, the value of Φ may change significantly. It is shown in [43] that the jump of AI during a single passage of the exact system through the vicinity of S is

$$\Delta\overline{\Phi} = \sqrt{\varepsilon}\,\overline{\Phi}\left(1 - \overline{\Phi}^2\right)^{1/4} f(\xi) \tag{1.11}$$

where $f(\xi)$ describes the dependence of the jump magnitude on the distance between the crossing point and the axis, parameterized by variable ξ, $\overline{\Phi} = \Phi/\Phi(0,0)$ is the normalized value of the AI. We refer the reader to [43] for the explicit definitions of $\xi, f(\xi)$, and $\Phi(0,0)$. The values of ξ and $\overline{\Phi}$ (and hence $\Delta\overline{\Phi}$) can be calculated exactly for any initial condition. However, a small change of order ε in the initial conditions produces, in general, a large (order 1) change in ξ. Hence, for

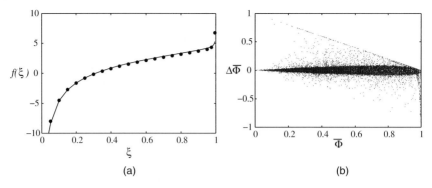

Figure 1.4 Jumps of AI: (a) the plot of $f(\xi)$. The solid line was obtained using analytical result (1.11) and the dots show the values obtained numerically from (1.10) for $\beta = 0.5$, $q = -1$, $\varepsilon = 10^{-3}$. (b) The distribution of the sizes of the jumps $\Delta\overline{\Phi}$ versus the values of $\overline{\Phi}$ before the crossings.

small ε it is possible to treat ξ as a random variable uniformly distributed on the unit interval [10].

Equation (1.11) was verified numerically for various values of parameters β, q, and ε. A typical plot of $f(\xi)$ is presented in Figure 1.4a. The function $f(\xi)$ – and hence $\Delta\overline{\Phi}$ – has singularities at both $\xi = 0$ and $\xi = 1$. Thus, there is a possibility (albeit quite small) of large changes in $\overline{\Phi}$ associated with a separatrix crossing. By direct calculation, the ensemble average of $\Delta\overline{\Phi}$ can be shown to vanish regardless of the value of $\overline{\Phi}$:

$$\langle \Delta\overline{\Phi} \rangle = \int_0^1 \Delta\overline{\Phi}(\xi, \overline{\Phi}) d\xi = 0 \tag{1.12}$$

1.3.3
Finite Perturbations

In most of the studies of resonance-induced chaotic diffusion, only infinitesimal perturbations were considered. To the best of our knowledge, the only papers that address, in any significant detail, the case of finite ε are [43, 53]. However, the dynamics in the presence of small but finite perturbations differs in several important ways from that with an infinitesimally small perturbation.

First, the very applicability of the method of averaging for larger ε is somewhat questionable as the ratio of the characteristic frequencies (e.g., $\Omega(\psi)$ and ε) may not be very large. Numerical simulations, however, indicate that the main result of the averaging method, that $\overline{\Phi}$ changes most significantly near the separatrix, holds for a wide range of ε. Moreover, this problem can be somewhat addressed by implementing the *improved adiabatic invariants*, see Section 1.4 below.

The second effect is that for finite values of ε, there is a finite probability that the jump size can become comparable to the range of the AI. Consequently, the boundaries of the system start playing an important role in the statistics of the jumps. While magnitudes of most of the jumps are still given by (1.11) and satisfy the zero-average statement, the distribution of large jumps differs from the original prediction. Indeed, (1.11) breaks down when the value of $\overline{\Phi}$ either before or after the separatrix crossing is close to one of the domain boundaries. Take, for example, $\varepsilon = 10^{-3}$ and $q = -1$. Then, approximately 0.1% of the jumps feel the presence of the boundaries. While the influence of the boundaries on the properties and statistics of single crossings (albeit for a different system) was discussed in details in [53], here we are interested in necessary modifications to the long-time dynamics of the system and, in particular, the rate of mixing.

Figure 1.4b presents the distribution of the sizes of the jumps $\Delta\overline{\Phi}$ versus the values of $\overline{\Phi}$ before the corresponding crossing. There are three types of jumps. Most of the jumps are small and concentrate near the $\overline{\Phi}$ axis (the densely covered region). These jumps are well described by (1.11). In particular, the average value of these jumps is zero. The second type are jumps corresponding to points that lie between the lines $\overline{\Phi}_{n+1} = \overline{\Phi}_n + \Delta\overline{\Phi}_n = \pm 1$, but outside of the densely covered region. These jumps happen when streamlines pass through the vicinities of the singularities of $\Delta\overline{\Phi}(\xi)$, given by (1.11). Such jumps were studied in detail in [53]. Finally, there are jumps that lie on either of the lines $\overline{\Phi}_{n+1} = \overline{\Phi}_n + \Delta\overline{\Phi}_n = 1$ and $\overline{\Phi}_n = 1$. They were called "axis crossings" in [55].

1.4
Passages Through Resonances in Autonomous Flows

In the current section, we will discuss another type of phenomena that occur at singular surfaces: scattering on and capture into resonance in 3D autonomous flows of the action–action–angle type. As an example, we consider a volume-preserving kinematic model inspired by a Stokes Taylor–Couette flow between two infinite counter-rotating coaxial cylinders (the "vertical" z axis is along the axis of the cylinders, ϱ is the distance from the axis, and θ is an angle in the "horizontal" plane, see [14] for a complete description). In the dimensionless units, the flow is

$$\begin{aligned}
\dot{\varrho} &= \varepsilon\kappa(\varrho-1)\cos\theta \\
\dot{z} &= \varepsilon(1+\ln\varrho/\ln\eta) \\
\dot{\theta} &= \omega(\varrho,z) - \frac{1}{\varrho}\varepsilon\kappa(2\varrho-1)\sin\theta
\end{aligned} \quad (1.13)$$

The value of ϱ changes between $\varrho = 1$ (at the inner cylinder) and $\varrho = 1/\eta$ (at the outer cylinder). The frequency of the unperturbed flow, $\omega(\varrho,z)$, is

$$\omega(\varrho,z) = -\varrho\frac{\eta}{1-\eta} + \frac{1}{\varrho}\frac{1}{1-\eta} + \frac{\eta}{1-\eta^2}\delta\sin(\lambda z)\left(\varrho - \frac{1}{\varrho}\right) \quad (1.14)$$

where $\lambda = 2\pi$ and δ are the wavenumber and amplitude of oscillations of the frequency of the outer cylinder, respectively. One can see that $\omega = 1$ and

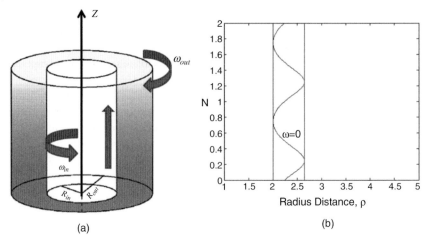

Figure 1.5 (a) The flow structure. (b) Division of the flow domain for z mod 1. A chaotic domain is between the vertical lines. A regular (KAM) domain consists of two parts at the left and at the right. The wavy line in the middle is the resonance, $\delta = 0.4$, $\eta = 0.2$.

$\omega = -1 + \delta \sin(\lambda z)$ on the inner and the outer cylinders, respectively. One can imagine the outer cylinder to consist of rings, each of which rotates with its own speed (see Figure 1.5a). The variables ϱ and z are the integrals of the unperturbed system. The unperturbed streamlines are circles with the direction of the rotation depending on the sign of $\omega(\varrho, z)$.

The perturbation consists of two parts. The first is a vertical (in the axial direction) shear-type flow (the $\dot{z}$ term). The second is an additional angular rotation due to a slight noncircularity of the outer cylinder. In (1.13), $0 \leq \varepsilon \ll 1$ is a small parameter, while $\kappa \sim 1$ defines a characteristic ratio of the two perturbations. The axial velocity, $\dot{z}$, equals ε at $\varrho = 1$ and vanishes at $\varrho = 1/\eta$.

For $\varepsilon > 0$, the variable θ is fast and the variables ϱ and z are slow. Thus, we can average (1.13) over one period of θ. The averaged trajectories (in the full 3D, (ϱ, z, θ), space) spiral around the cylinders of constant radius ($\varrho = $ const) with the direction of the rotation depending on the sign of ω. The quantity

$$\Phi = \varrho$$

is an integral of the averaged system and is an AI of the exact system. The averaging is valid away from a resonance surface (in 3D), or a curve on the slow, (ϱ, z)-plane where $\omega = 0$. We denote that surface by R. It follows from (1.14) that R is given by

$$\varrho_R^2(z) = \frac{1}{\eta} \frac{1 + \eta - \eta \delta \sin(\lambda z)}{1 + \eta - \delta \sin(\lambda z)}$$

and located between $\varrho_{\min}$ and $\varrho_{\max}$. The division of the flow domain is shown in Figure 1.5b (note that we plotted z mod 1). Trajectories to the left and to the right of the corresponding vertical lines do not cross R.

As a passive tracer approaches R, it can be either *scattered at a resonance* or *captured into resonance*. While the scattering on resonance is somewhat similar to what happens at the separatrix crossing, capture can occur only at a resonance. Qualitatively, the difference between the two regimes can be described as follows. In the case of capture, upon arrival into the resonant zone, the phase switches its behavior from rotation to oscillation. The system drifts along the resonant surface for a long, of order ε^{-1}, time. As a result, the value of the AI changes by $\Delta\Phi = O(1)$. Among all the streamlines that arrive to the resonant zone during a given time interval (of order ε^{-1}), only a small, $O(\sqrt{\varepsilon})$ part of streamlines are captured. In the case of scattering there is no phase oscillation. The streamlines pass through the resonance zone in an $O(\sqrt{\varepsilon})$ time and the corresponding jump in the AI is $\Delta\Phi = O(\sqrt{\varepsilon})$. We describe these two processes below.

1.4.1
Scattering on Resonance

For most initial conditions, a tracer passes through the vicinity of R in a relatively short time and without capture. In such a case, in the first approximation we can fix the value of slow variables ω and z at the resonance values, and the dynamics is defined in terms of a forced-pendulum type of the second-order ODE for θ:

$$\theta'' = a + b_1 \cos\theta \tag{1.15}$$

where

$$a = \frac{\eta}{1-\eta^2} \delta\lambda \cos(\lambda z)\left(\varrho - \frac{1}{\varrho}\right)(1 + \ln\varrho/\ln\eta) \quad b_1 = -2\kappa \frac{1}{\varrho+1}$$

and the prime denotes the derivative with respect to the rescaled time $\bar{t} = \sqrt{\varepsilon}\,t$. System (1.15) can be described by the resonance potential $V = -a\theta - b_1 \sin\theta$. The shape of phase portraits for the motion in the potential V depends on the relation between a and b_1. If

$$|b_1| > |a| \tag{1.16}$$

the phase portrait looks like the one shown in Figure 1.6a, and vice versa for Figure 1.6b.

In the process of scattering, the value of Φ undergoes a jump, the magnitude of which is (in the main approximation) given by

$$\Delta\Phi = -2s\sqrt{\varepsilon}\kappa\,\frac{\varrho-1}{\sqrt{|a|}}\int_{s\infty}^{\bar{\theta}_*} \cos\theta\sqrt{2|s2\pi\xi + \theta + (b_1/a)\sin\theta|}\,d\theta \tag{1.17}$$

where $\bar{\theta}_*$ is the value of θ at the crossing, $s = \mathrm{sign}(a)$, and $\xi = \{V(\theta_*)/(2\pi|a|)\} \in (0,1)$, where the curly brackets denote the fractional part. If (1.16) holds, the ensemble average of $\Delta\Phi$ is

$$\langle\Delta\Phi\rangle = -s\sqrt{\varepsilon}\,\frac{\varrho^2-1}{2\pi}\,S_R$$

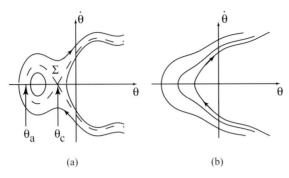

Figure 1.6 Schematic phase portraits on the $(\theta,\dot\theta)$ plane.

Here S_R is the area under the separatrix loop, Σ, in Figure 1.6a:

$$S_R = 2\left|\int_{\theta_a}^{\theta_c} \sqrt{-2(V-V_c)}d\theta\right|$$

where V_c is the value of V at the hyperbolic fixed point in Figure 1.6a. If (1.16) does not hold, $\langle\Delta\Phi\rangle = 0$, as there is no separatrix, $S_R = 0$.

Equation (1.17) was checked numerically for various values of parameters ξ, κ, and ε. In Figure 1.7, the plots of $\Delta\varrho(\xi)/\sqrt{\varepsilon}$ are presented for (a) $\kappa = 2$ (when (1.16) is satisfied) and (b) $\kappa = 0.2$ (when (1.16) is not satisfied). The solid lines in Figure 1.7 correspond to theoretical values of $\Delta\varrho(\varepsilon)/\sqrt{\varepsilon}$ and the asterisks show values obtained numerically from (1.13) for various values of ξ. When (1.16) is satisfied, $\Delta\varrho(\xi)$ has a singularity.

1.4.2
Capture Into Resonance

The other phenomenon that affects the behavior of streamlines at a resonance crossing is capture into resonance. See [14, 56] for additional details. Capture is possible only if the phase portrait in the $(\theta,\dot\theta)$-plane looks like the one shown in

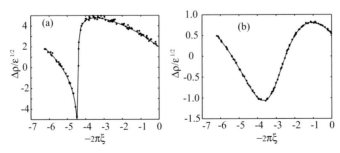

Figure 1.7 The plot of $\Delta\varrho/\sqrt{\varepsilon}$ as a function of ξ; (a) $\kappa=2$ and (b) $\kappa=0.2$. Note the difference in scales.

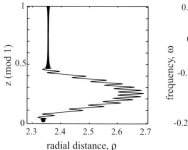
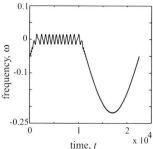

Figure 1.8 Captured motion. (a) A projection of a streamline on the slow, (ϱ, z), plane; and (b) the time evolution of $\omega(\varrho, z)$, $\varepsilon = 10^{-4}$, $\kappa = 2$.

Figure 1.6a, in other words, if there is a separatrix in the $(\theta, \dot{\theta})$-plane. Let $\Pi(\varrho_*, z_*)$ be a flux of resonance flow through the separatrix loop in Figure 1.6a. The value of $\Pi(\varrho_*, z_*)$ changes as a phase point moves along a streamline. If $\Pi(\varrho_*, z_*)$ is decreasing, the capture is not possible. If Π increases and a streamline comes very close to the hyperbolic fixed point, it may cross Σ and, as a result, be caught in the oscillatory domain within the separatrix loop. In this case, a streamline starts shadowing the resonant surface. The captured motion is integrable and Hamiltonian. Depending on the structure of resonance, a tracer can be released from resonance (which is the case in the system under consideration) or reach the boundary of the system.

The dynamics of a typical capture is shown in Figure 1.8 as a projection on the slow, (ϱ, z), plane and the time evolution of $\omega(\varrho, z)$. A streamline comes from the bottom in Figure 1.8a (from the left in Figure 1.8b), is captured near $z = 0.05$ ($t = 100$), moves along the resonance, is released from the resonance near $z = 0.45$ ($t = 1000$), and then proceeds along an adiabatic path.

As it was discussed in [10, 13, 39], capture can be considered as a probabilistic phenomenon: initial conditions for streamlines that are or are not captured are mixed. The probability of capture for the streamlines starting inside a small ball centered at a certain point can be defined as a ration of the measure of the initial conditions that are captured to the full measure of the ball. It was proved in [10] that this probability is of the order of $O(\sqrt{\varepsilon})$.

1.4.3
Improved AI

As the value of ε increases, it becomes more and more difficult to distinguish between the jumps of AI in the process if scattering, which are of the order $O(\sqrt{\varepsilon})$ from oscillations of AI away from R, which are of the order $O(\varepsilon)$. In this situation the notion of *improved* AI becomes very useful. The improved AI $\Phi^{(1)}$ is given by

$$\Phi^{(1)} = \varrho - \varepsilon \kappa \frac{\varrho - 1}{\omega(z, \varrho)} \sin \theta$$

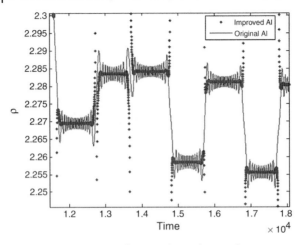

Figure 1.9 Comparison of improved AI with original AI (ϱ).

It can be checked by the direct calculations that oscillations of $\Phi^{(1)}$ away from the resonance are of the order of ε^2, while its jumps due to scattering or still $O(\sqrt{\varepsilon})$. However, there is no such a thing as free lunch: improved AI has a singularity at the resonance (which is natural as it contains ω in the denominator of the correction term), while the original AI remains finite, see Figure 1.9.

1.4.4
Jump of AI Between First- and Second-Layer Boundaries

The numerical simulations show that there is large transport of particles between the domain inside first-layer boundaries and the domain outside first-layer boundaries. We find out that the jump of the AI also happens for particles in the domain outside first-layer boundaries, where the resonance does not exist. The jump of AI happens at $z_{in} = 3/4$ and $z_{out} = 1/4$ (near the inner and outer first-layer boundary, respectively). Thus, the jump happens only once per slow period in the domain outside first-layer boundaries, instead of twice per slow period in the domain between first-layer boundaries.

The magnitude of $\Delta\varrho$ outside first-layer boundaries depends on both the value of θ at the point of closest approach and the distance between the particle and the first-layer boundary, shown in Figure 1.10a. When particles are close to the first-layer boundaries, the magnitude of $\Delta\varrho$ is large, and ω is relatively small (of order of $\sqrt{\varepsilon}$) at z_{in} close to the inner first-layer boundary, and at z_{out} close to the outer first-layer boundary. As particles move further from the first-layer boundaries, ω increases and the magnitude of $\Delta\varrho$ decreases. Figure 1.10b represents the variance of the distribution of $\Delta\varrho$ for uniformly distributed θ as the function of ϱ. The absolute values of $\Delta\varrho$ and the variance of the distribution of $\Delta\varrho$ decrease as the distance between particles and the first-layer boundaries increases. We can use the variance of $\Delta\varrho$ to estimate the approximate position of the boundaries which the streamlines that

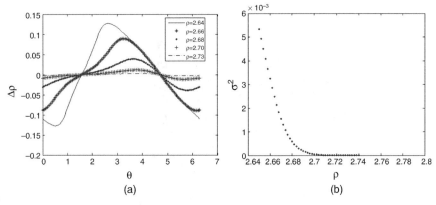

Figure 1.10 (a) The plot of $\Delta\varrho$ as a function of v for different values of ϱ. (b) The variance of distribution of $\Delta\varrho$ as a function of ϱ.

start in the mixing domain do not penetrate. The second-layer boundaries are the last invariant tori, and they are the actual boundaries of the chaotic domain.

1.4.5
Long-Time Dynamics and Adiabatic Diffusion

In many systems with random walks, the spreading of trajectories (or streamlines) can be described using diffusion-type equation(1.6). The coefficients, $U(\varrho)$ and $D(\varrho)$, are

$$U_A(\varrho) = \langle \Delta\varrho \rangle \quad D(\varrho) = \int_0^1 (\Delta\varrho(\xi) - \langle \Delta\varrho \rangle)^2 d\xi \tag{1.18}$$

As it was discussed above (and also illustrated below), for very small values of ε, U_A, averaged over a slow period, very often vanishes everywhere in the flow domain except in the very vicinity of the boundaries. The boundary conditions for solving diffusion equation (1.6) are von Neumann (no flux) at the outer boundary of the second-layer domain.

To study the validity of the adiabatic diffusion approximation, we performed a set of numerical simulations using the values of parameters specified above and $\varepsilon = 10^{-3}$, $\kappa = 0.2$. We choose 1000 initial uniformly distributed in a small cubic box in the size of ε picked from the middle of the chaotic domain. Its size was $\varrho_{in} \times z_{in} \times \theta_{in} = [2.220, 2.229] \times [0.251, 0.260] \times [0.011, 0, 020]$. We considered the Poincaré sections located at $z = N + 0.25$ and $z = N + 0.75$, where N is a set of integer numbers. Every trajectory crosses the resonance once between two consecutive sections.

The results of our numerical simulations showed that the particles which initially concentrated in those small boxes start to diffuse after multiple resonance crossings, and in the end, the distribution of particles in the radius direction is quite uniform, shown in Figure 1.11. The solid line in Figure 1.11 is the solution of diffusion

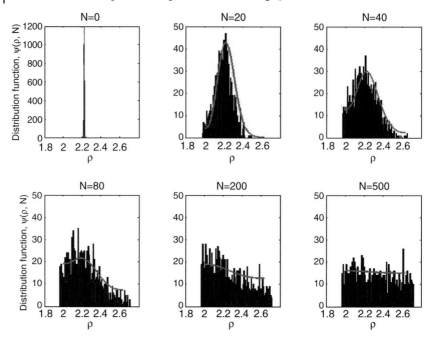

Figure 1.11 The histogram of $P(\varrho, N)$ after different numbers of resonance crossings for Box 1.

equation (1.6) using the said box as the initial condition, and both results are consistent with each other. The same results are also obtained for other initial conditions, located in different parts of the flow domain.

The second moment of the distribution function $P(\varrho, N)$ of numerical simulations, σ^2, is shown as the dash line in Figure 1.12. The constant slope of the solid line in Figure 1.12 is the diffusion coefficient $D(\varrho)$ analytically calculated using Eq. (1.18) for ($\langle \varrho_0 \rangle = 2.225$) (in the center of the box). The second moment of numerical simulations σ^2 is very close to $D(\varrho)$ in the beginning, before particles reaching the first-layer boundaries. However, when particles start to cross the first-layer boundaries, σ^2 and $D(\varrho)$ start to diverge. In the end, σ^2 comes to an asymptotic value with small oscillations. That asymptotic value is the variance of uniformly distributed particles in the chaotic domain, shown as horizontal lines in Figure 1.12.

1.5
Passages Through Resonances in Nonautonomous Flows

The effect of resonances is quite richer in the multifrequency systems. An example of such systems is nonautonomous flows, where the time appears explicitly as an additional fast phase. In the present section we follow [4] and discuss the mixing dynamics in such systems and, as an example, we consider an incompressible fluid flow in a one-dimensional array of cubic cells described by the following equations:

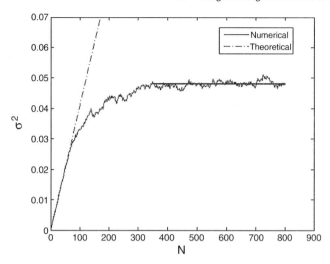

Figure 1.12 The variance of ϱ, over 1000 trajectories as a function of the number of resonance crossings for Box 1.

$$\begin{aligned}
\dot{x} &= -\cos(\pi x)\sin(\pi y) + \pi b \sin(\pi x)\sin(\pi y)\sin\omega t \\
&\quad + \varepsilon \sin(2\pi x)\sin(\pi z) \\
\dot{y} &= \sin(\pi x)\cos(\pi y) + \pi b \cos(\pi x)\cos(\pi y)\sin\omega t \\
&\quad + \varepsilon \sin(2\pi y)\sin(\pi z) \\
\dot{z} &= 2\varepsilon \cos(\pi z)[\cos(2\pi x) + \cos(2\pi y)]
\end{aligned} \quad (1.19)$$

This is a linearization of a system introduced by Solomon and Mezic in [57] as a qualitative model of a Lorenz-force driven cellular flow in a channel of rectangular cross-section $(-0.5 < y, z < 0.5)$. It is easy to check that the no-slip boundary condition at the channel walls is not satisfied. However, the exact solution of the Navier–Stokes equations satisfying the proper boundary conditions will lead us to qualitatively the same conclusions while making the calculations unnecessarily complicated. The terms proportional to ε describe a weak correction to the main recirculation flow caused by inertial effects (Eckman pumping). The time dependence of the flow represents an external perturbation describing the shift, with amplitude b, of the boundaries between the cells (planes $x = n + 1/2$, $n \in \mathbb{Z}$). For nonzero b, there is transport between the cells; however, our objective here is to understand the transport properties of the flow inside each of the cells. Since the dynamics in all the cells are identical, we will consider only the cell with $-0.5 < x < 0.5$.

1.5.1
Unperturbed Flow

First, consider the unperturbed (base) flow characterized by $\varepsilon = 0$ and $b = 0$. In this case, (1.19) is reduced to

$$\dot{x} = -\cos(\pi x)\sin(\pi y) \quad \dot{y} = \sin(\pi x)\cos(\pi y) \quad \dot{z} = 0 \quad (1.20)$$

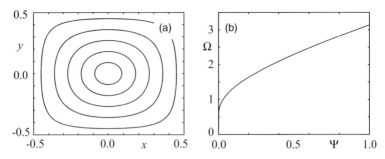

Figure 1.13 Unperturbed system: (a) typical streamlines in the $z = $ const plane and (b) the frequency $\Omega(\Psi)$.

This is a two-dimensional autonomous flow which possesses two invariants:

$$z = \text{const} \quad \Psi = \cos(\pi x)\cos(\pi y) = \text{const}$$

with Ψ proportional to the streamfunction of the unperturbed flow in the (x, y) plane. All the streamlines $\Gamma_{z,\Psi}$ of the unperturbed flow are closed (see Figure 1.13a) with the period of motion $T(\Psi)$. The corresponding frequency $\Omega = 2\pi/T$ ranges from $\Omega = 0$ at the boundaries of the cell to $\Omega = \pi$ in the center. On every $\Gamma_{z,\Psi}$, we can introduce a uniform phase χ mod (2π) such that $\chi = 0$ on the positive x-axis and $\dot{\chi} = \Omega(\Psi)$. Due to the dependence of Ω of Ψ, the unperturbed flow is characterized by mixing in only one dimension (along the streamlines $\Gamma_{z,\Psi}$) at $O(1)$ rate.

1.5.2
Two Perturbations and Averaging

Next, consider the effect of the Eckman pumping ($\varepsilon > 0$), ignoring the time-dependent shift for the moment ($b = 0$). In this limit, flow (1.19) is steady but conserves neither z nor Ψ. The dynamics is characterized by two different timescales: the variable χ is fast (changes on $O(1)$ timescale), while the variables z and Ψ are slow (change on $O(\varepsilon^{-1})$ timescale). Averaging evolution equations for $\dot{\Psi}$ and $\dot{z}$ over the quick oscillations in χ over one period of the unperturbed motion, we obtain an averaged system that possesses an AI Φ defined as a flux of the vector field v_ε (the ε-dependent part of the perturbation in (1.19), through a surface S_Γ bounded by an unperturbed streamline $\Gamma_{z,\Psi}$). The averaged system can be written as

$$\dot{\Psi} = -\varepsilon \frac{\pi}{T(\Psi)} \frac{\partial \Phi}{\partial z} \quad \dot{z} = \varepsilon \frac{\pi}{T(\Psi)} \frac{\partial \Phi}{\partial \Psi} \tag{1.21}$$

The addition of the time-dependent perturbation makes the structure of the flow much more complex if b and ε are of similar magnitude. Hence, in what follows we assume $\beta \equiv b/\varepsilon = O(1)$. Furthermore, we assume $\omega = O(\Omega) = O(1)$. The evolution equations for the slow variables are

$$\dot{\Psi} = -2\pi\varepsilon \sin(\pi z)\Psi\left(\sin^2(\pi x)+\sin^2(\pi y)\right)-\varepsilon\frac{\pi^2}{2}\beta(\sin(2\pi y)\sin(\omega t))$$
$$\dot{z} = 2\varepsilon \cos(\pi z)[\cos(2\pi x)+\cos(2\pi y)]$$
(1.22)

If Ω and ω are incommensurate, then averaging over χ and t can be performed independently (see, e.g., [58]). In this case, the time-dependent terms in the equation for $\dot{\Psi}$ average out and we would expect the AI Φ to be conserved as before. The evolution over a longer time interval shows that the AI remains essentially constant except for the short periods of time when $\Omega \approx \omega$, as Figure 1.14b illustrates. We therefore find, as previous studies [7, 59–61] did, a clear indication of the fact that the breakdown of adiabatic invariance is a consequence of processes occurring in the vicinity of resonances.

1.5.3
Resonant Phenomena

As the value of Ψ slowly drifts, so does $\Omega(\Psi)$. Hence, at certain values of Ψ a resonance condition

$$n\Omega(\Psi)-\omega = 0 \qquad (1.23)$$

is satisfied for some nonzero integer n. Note that (1.23) is a special case of a more general resonance condition $n\Omega = m\omega$, which corresponds to a generic time-periodic perturbation. The restriction to $m=1$ in our case is a consequence of the particular form of the time-dependent perturbation in (1.19), namely, that only the first harmonic is present. Since Ω is independent of z, all resonance surfaces R_n, defined by $\Psi(x,y) = \Psi_n \equiv \Omega^{-1}(\omega/n)$, are vertical cylinders in the physical space or vertical lines $\Psi = \Psi_n$ in the slow plane (see Figure 1.14b).

In what follows we describe a quantitative description of the dynamics near resonant surfaces based on the theory of resonance phenomena in multiple-frequency systems [58]. Note that, unlike autonomous systems with resonance phenomena (see, e.g., [14], and Section 1.4) where the fast (unperturbed) dynamics slows

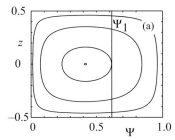
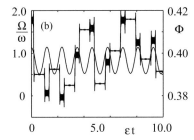

Figure 1.14 The complete flow with $\varepsilon = b = 10^{-4}$ and $\omega = 2.5$: (a) projection of the averaged system on the slow plane, the bold line is the 1 : 1 resonance is shown; (b) plots of Ω/ω (dashed line) and Φ (solid line) versus time for complete flow (1.1).

down near a resonance, for the flow studied here, near R_n both the fast angle variable χ and the phase of the perturbation ωt keep changing rapidly. It is only a particular linear combination of these phases that slows down:

$$\gamma = n\chi - \omega t \tag{1.24}$$

Therefore, near every resonance surface there is just one fast variable, χ, rather than two far from the resonances. Of the three other variables, Ω and z are slow and γ is semislow with characteristic rates of change of order $O(\varepsilon)$ and $O(\varepsilon^{1/2})$, respectively. We can still average the exact equations of motion for all three slow and semislow variables over χ (perform the so-called *partial averaging*, see [58]) in order to obtain equations of motion near a resonance surface:

$$\gamma' = \frac{1}{\sqrt{\varepsilon}}(n\Omega - \omega) \quad \Omega' = \sqrt{\varepsilon}\frac{\partial \Omega}{\partial \Psi}\frac{\dot{\Psi}}{\varepsilon} \quad z' = \sqrt{\varepsilon}\frac{\dot{z}}{\varepsilon} \tag{1.25}$$

In (1.25), the prime denotes the derivative with respect to the rescaled time $\bar{t} = \sqrt{\varepsilon}\,t$ and $\dot{\Psi}$, $\dot{z}$ were defined in (1.22). Equation (1.25) is very similar to the one appearing in autonomous case (see (1.15)).

For most initial conditions, tracers pass through the vicinity of resonance in a relatively short time and the value of Φ undergoes a relatively small jump. In the first approximation we can fix the value of slow variables Ω and z at the resonance values, which yields a forced pendulum-like equation for γ:

$$\gamma'' = \frac{1}{\sqrt{\varepsilon}}n\Omega' = a_n + b_n \cos \gamma \tag{1.26}$$

In (1.26), a_n and b_n correspond to the average of the first and the second term in (1.22) over the fast period $T_n \equiv T(\Psi_n) = 2\pi n/\omega$, respectively:

$$a_n = -n\Psi \sin(\pi z_*) \frac{\partial \Omega}{\partial \Psi} \int_0^{2\pi} (\sin^2(\pi x) + \sin^2(\pi y)) d\chi$$

$$b_n = -\frac{\pi}{4} n\beta \frac{\partial \Omega}{\partial \Psi} \int_0^{2\pi} \sin(2\pi y) \sin(n\chi) d\chi$$

and z_* is the value of z at which the crossing occurs. The average of $\dot{\Phi}$ over T_n can be computed using (1.21) and (1.22), yielding

$$\langle \dot{\Phi} \rangle = \varepsilon \pi \beta \, c_n \cos(\pi z_*) \cos \gamma \tag{1.27}$$

where the coefficient c_n is given by

$$c_n = -\frac{n}{\omega} \int_0^{2\pi} (\cos(2\pi x) + \cos(2\pi y)) \sin(2\pi y) \sin(n\chi) d\chi.$$

Integrating over a time interval during which the resonance is crossed once, we obtain in the leading order

$$\Delta\Phi = -2\sqrt{\varepsilon}\,s\pi\beta\cos(\pi z_*)\frac{c_n}{\sqrt{|a_n|}}\int_{-s\infty}^{\gamma(t_*)}\frac{\cos\gamma}{\sqrt{2|s2\pi\xi+\gamma+(b_n/a_n)\sin\gamma|}}d\gamma$$

where ξ is defined analogously to the previous section.

When $|b_n| > |a_n|$, $\langle\Delta\Phi\rangle$ is finite:

$$\langle\Delta\Phi\rangle = -s\sqrt{\varepsilon}\,\beta\cos(\pi z_*)\frac{c_n}{b_n}S$$

where S is the area under the separatrix loop on the resonance plane (cf. Figure 1.6a). In the opposite case $|b_n| < |a_n|$, there is no separatrix, $S = 0$, and hence $\langle\Delta\Phi\rangle = 0$. Generally, a nonzero ensemble average of $\Delta\Phi$ results in a unidirectional drift of Φ. However, in the current problem, two successive crossings occur at almost the opposite values of z. Thus, they cancel each other *on average* because of the change of the sign of s, and the aggregate change of Φ on one period of the slow motion has zero mean. The second moment of $\Delta\Phi$

$$\sigma^2 = \int_0^1 (\Delta\Phi(\xi) - \langle\Delta\Phi\rangle)^2 d\xi,$$

is finite for any value of Φ. The dependence of $\Delta\Phi$ on the order of the resonance is determined by the scaling of a_n, b_n, and c_n, which are the Fourier coefficients of smooth functions. In particular, a_n corresponds to the 0-th harmonic and increases linearly with n. On the other hand, b_n and c_n correspond to higher harmonics and decrease exponentially with n. As a consequence, the characteristic magnitude of the jumps decays exponentially. Thus, only low-order resonances contribute significantly to the change in the value of the AI. For high-order resonances, we have $\Delta\Phi \sim \sqrt{\varepsilon}\,e^{-an}$, where a is some constant.

1.5.4
Volume of the Mixing Domain

On every period $T_\varepsilon(\Phi)$ of the slow motion along a given trajectory, the value of Ψ changes between $\Psi_{\min}$ and $\Psi_{\max}$. If no (low-order) resonance Ψ_n falls in this interval, then that trajectory (and all trajectories inside of it) remains regular. If, on the other hand, the trajectory crosses a resonance surface, the AI experiences jumps and the motion becomes chaotic.

In the $\varepsilon \to 0$ limit, the boundary between the chaotic and the regular domains is, thus, given by the trajectory Γ_{Φ^*} that (i) touches a resonance surface and (ii) has the largest value Φ among all such trajectories on the (Ψ, z) plane (bold line in Figure 1.14a). Condition (ii) is necessary when multiple resonances are considered. In the physical space, the boundary is formed by the corresponding torus τ_{Φ^*}. The Poincaré section of the complete flow by the plane $z = 0$ (see Figure 1.15a) confirms that the space inside the torus τ_{Φ^*} corresponds to the regular domain discovered in [57], while the rest of the physical space belongs to the chaotic domain. Moving the frequency ω closer to the resonance with $\Omega(\Psi_c) \approx 2.2$ completely wipes out the regular domain (see Figure 1.15b).

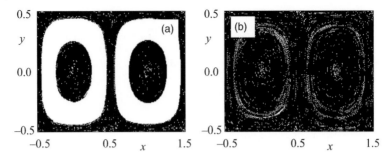

Figure 1.15 Complete and partial mixing. $z = 0$ Poincaré section of a single streamline with $\varepsilon = b = 10^{-4}$ and (a) $\omega = 4.0$, (b) $\omega = 2.5$.

The width d of the regular domain can be computed easily for any value of ω (see Figure 1.16). For $0 < \omega \ll 1$, all the resonances are located near $\Psi = 0$ (i.e., cell boundary). As ω is increased, the 1 : 1 resonance is the first to penetrate deeper into the cell. For $0 < \omega \pi$, Γ_{Φ^*} is tangent to the resonance $\Psi = \Psi_1$ (see Figure 1.14a). As ω approaches π, the 1 : 1 resonance is pushed out of the cell and the 1 : 3 resonance becomes the most prominent for $\pi \omega 3\pi$ (recall that even resonances do not lead to jumps in Φ and thus do not contribute to adiabatic diffusion). Then the process repeats itself: as ω is increased further, low-order resonances are pushed out of the cell and higher resonances become prominent. Finally, as $\omega \to \infty$, the cell becomes uniformly covered by high-order resonances. However, the impact of the high resonances is exponentially small and hence we can expect mixing to become spatially uniform on exponentially long timescales. On finite timescales characteristic of experiments (e.g., those reported in [57]), it would appear that no mixing is taking place. We should also point out that, for the flow between concentric spheres considered in [60, 61], complete mixing relies on high-order resonances and, while conceptually possible, would similarly require exceedingly long times.

In our case, complete mixing on experimentally accessible timescales can be achieved by eliminating the domain of regular dynamics via a proper placement of a low-order resonance. This can be accomplished by setting the frequency ω of the

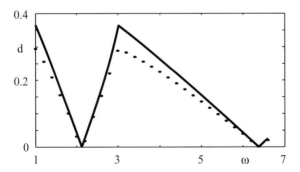

Figure 1.16 The width d of the regular domain as a function of the perturbation frequency, ω for $\varepsilon = b = 10^{-4}$: solid line – theoretical prediction, dots – numerical simulations.

perturbation such that $\Psi_n(\omega) = \Psi_c$ for some n. More precisely, the resonance must be within the interval $|\Psi_n - \Psi_c| = O(\sqrt{\varepsilon})$, as the chaotic domain penetrates inside Γ_{Φ^*} by an $O(\sqrt{\varepsilon})$ distance [50]. This property, negligible in most similar problems, is important here as the magnitude of the jumps vanishes at Ψ_c. Indeed, $\dot{\Phi} \sim \dot{z}$ according to (1.27), so $\Delta\Phi = 0$ at $\Psi = \Psi_c$, as $\dot{z} = 0$ there. Since the width of the regular domain $d \sim |\Psi_n(\omega) - \Psi_c|$, we find the width of the frequency intervals yielding complete mixing (where $d \approx 0$) to scale as $\Delta\omega \sim \sqrt{\varepsilon}$ (see Figure 1.16).

Acknowledgments

This work was supported in part by the National Science Foundation (project CBET-0900177) and the Russian Foundation for Basic Research (projects 09-01-00333, 08-02-00201). The author is grateful to his co-authors Roman Grigoriev, Igor Mezic, Anatoly Neishtadt, Alexei Vasiliev, and John Widloski.

References

1 Mezic, I. and Wiggins, S. (1994) On the integrability and perturbation of three-dimensional fluid flows with symmetry. *Journal of Nonlinear Science*, **4**, 157–194.
2 Haller, G. and Mezic, I. (1998) Reduction of three-dimensional, volume preserving flows by symmetry. *Nonlinearity*, **11**, 319.
3 Grigoriev, R.O. (2005) Chaotic mixing in thermocapillary-driven microdroplets. *Physics of Fluids*, **17**, 033601.
4 Vainchtein, D.L., Widloski, J., and Grigoriev, R.O. (2008) Resonant mixing in perturbed action–action–angle flow. *Physical Review E*, **78**, art: 026302.
5 Laskar, J. (1996) Large scale chaos and marginal stability in the solar system. *Celestial Mechanics & Dynamical Astronomy*, **64**, 115–162.
6 Dellnitz, M., Junge, O., Koon, W.S., Lekien, F., Lo, M.W., Marsden, J.E., Padberg, K., Preis, R., Ross, S.D., and Thiere, B. (2005) Transport in dynamical astronomy and multibody problems. *International Journal of Bifurcation and Chaos*, **15**, 699–727.
7 Feingold, M., Kadanoff, L.P., and Piro, O. (1988) Passive scalars, 3-dimensional volume-preserving maps, and chaos. *Journal of Statistical Physics*, **50**, 529–565.
8 Arnold, V.I., Kozlov, V.V., and Neishtadt, A.I. (1988) *Dynamical Systems III*.

Encyclopedia of Mathematical Sciences, Springer, New York, NY.
9 Vainshtein, D.L., Vasiliev, A.A., and Neishtadt, A.I. (1996) Changes in the adiabatic invariant and streamline chaos in confined incompressible Stokes flow. *Chaos*, **6**, 67–77.
10 Neishtadt, A.I. (1999) On adiabatic invariance in two-frequency systems. in: Hamiltonian systems with 3 or more degrees of freedom. *NATO ASI Series C*, **533**, 193–213.
11 Chirikov, B.V. (1959) Passage of nonlinear oscillatory system through resonance. *Soviet Physics Doklady*, **4**, 390–394.
12 Kevorkian, J. (1974) Model for reentry roll resonance. *SIAM Journal on Applied Mathematics*, **26**, 638–669.
13 Neishtadt, A.I. (1997) Scattering by resonances. *Celestial Mechanics & Dynamical Astronomy*, **65**, 1–20.
14 Vainchtein, D.L., Neishtadt, A.I., and Mezic, I. (2006) On passage through resonances in volume-preserving systems. *Chaos*, **16**, 043123.
15 Neishtadt, A.I. (1986) Change of an adiabatic invariant at a separatrix. *Soviet Journal of Plasma Physics*, **12**, 568–573.
16 Cary, J.R., Escande, D.F., and Tennyson, J.L. (1986) Adiabatic-invariant change due

to separatrix crossing. *Physical Review A*, **34**, 4256–4275.

17 Haberman, R. and Bourland, F.J. (1994) Slow passage through homoclinic orbits for the unfolding of a saddle-center bifurcation and the change in adiabatic invariant. *Studies in Applied Mathematics*, **91**, 95–124.

18 Vainshtein, D.L., Zelenyi, L.M., Neishtadt, A.I., and Savenkov, B.V. (1999) Jumps in an adiabatic invariant with small initial values. *Plasma Physics Reports*, **25**, 333–337.

19 Diminnie, D.C. and Haberman, R. (2002) Connection across a separatrix with dissipation. *Physica D*, **162**, 34–52.

20 Neishtadt, A.I. and Vasiliev, A.A. (1999) Change of the adiabatic invariant at a separatrix in a volume-preserving 3D system. *Nonlinearity*, **12**, 303–320.

21 Ward, T. and Homsy, G.M. (2003) Electrohydrodynamically driven chaotic mixing in a translating drop. II. Experiments. *Physics of Fluids*, **15**, 2987–2994.

22 Grigoriev, R.O., Schatz, M.F., and Sharma, V. (2006) Chaotic mixing in microdroplets. *Lab Chip*, **6**, 1369–1372.

23 Paik, P., Pamula, V.K., Pollack, M.G., and Fair, R.B. (2003) Rapid droplet mixers for digital microfluidic systems. *Lab Chip*, **4**, 253–259.

24 Fowler, J., Moon, H., and Kim, C.-J. (2002) Enhancement of mixing by droplet-based microfluidics. in Proc. IEEE Conf. MEMS, pp. 97–100.

25 Song, H., Tice, J.D., and Ismagilov, R.F. (2003) A microfluidic system for controlling reaction networks in time. *Angewandte Chemie-International Edition*, **42**, 768–772.

26 Belbruno, E. and Marsden, B.G. (1997) Resonance hopping in comets. *Astronomical Journal*, **113**, 1433–1444.

27 Jaffe, C., Ross, S.D., Lo, M.W., Marsden, J., Farrelly, D., and Uzer, T. (2002) Statistical theory of asteroid escape rates. *Physical Review Letters*, **89**, 011101.

28 Neishtadt, A.I. and Sidorenko, V.V. (2004) Wisdom system: Dynamics in the adiabatic approximation. *Celestial Mechanics & Dynamical Astronomy*, **90**, 307–330.

29 Zabiego, M., Ghendrih, P., Becoulet, M., Costanzo, L., De Michelis, C., Friant, C., and Gunn, J. (2001) Characterisation of the separatrix position in the ergodic divertor discharges of the tore supra tokamak. *Journal of Nuclear Materials*, **290**, 985–989.

30 Lehnen, M. et al. (2005) Transport and divertor properties of the dynamic ergodic divertor. *Plasma Physics and Controlled Fusion*, **47**, B237–B248.

31 Jakubowski, M.W. et al. (2006) Change of the magnetic-field topology by an ergodic divertor and the effect on the plasma structure and transport. *Physical Review Letters*, **96**, 035004.

32 Gendelman, O., Manevitch, L.I., Vakakis, A.F., and M'Closkey, R. (2001) Energy pumping in nonlinear mechanical oscillators: Part I – Dynamics of the underlying Hamiltonian systems. *Journal of Applied Mechanics – Transactions of the ASME*, **68**, 34–41.

33 Vakakis, A.F. and Gendelman, O. (2001) Energy pumping in nonlinear mechanical oscillators: Part II – resonance capture. *Journal of Applied Mechanics-Transactions of the ASME*, **68**, 42–48.

34 Popov, A.A., Thompson, J.M.T., and McRobie, F.A. (2001) Chaotic energy exchange through auto-parametric resonance in cylindrical shells. *Journal of Sound and Vibration*, **248**, 395–411.

35 Itin, A.P., Neishtadt, A.I., and Vasiliev, A.A. (2001) Resonant phenomena in slowly perturbed rectangular billiards. *Physics Letters A*, **291**, 133–138.

36 Chernikov, A.A. and Schmidt, G. (1994) Adiabatic chaos in Josephson-junction arrays. *Physical Review E*, **50**, 3436–3445.

37 Vasiliev, A.A., Neishtadt, A.I., and Itin, A.P. (1997) On dynamics of four phase-coupled oscillators with close frequencies. *Regular and Chaotic Dynamics*, **3**, 1–9.

38 Ashour-Abdalla, M., Frank, L., Paterson, M., Peroomian, V., and Zelenyi, L.M. (1996) Proton velocity distributions in the magnetotail: Theory and observations. *Journal of Geophysical Research*, **101**, 2587–2598.

39 Itin, A.P., Neishtadt, A.I., and Vasiliev, A.A. (2000) Captures into resonance and scattering on resonance in dynamics of a charged relativistic particle in magnetic

field and electrostatic wave. *Physica D*, **141**, 281–296.

40 Vainchtein, D.L., Rovinsky, E.V., Zelenyi, L.M., and Neishtadt, A.I. (2004) Resonances and particle stochastization in nonhomogeneous electromagnetic fields. *Journal of Nonlinear Science*, **14**, 173–205.

41 Vainchtein, D.L., Vasiliev, A.A., and Neishtadt, A.I. (2009) Electron dynamics in a parabolic magnetic field in the presence of an electrostatic wave. *Plasma Physics Reports*, **35** (12), 1021–1031.

42 Neishtadt, A.I., Artemyev, A.V., Zelenyi, L.M., and Vainshtein, D.L. (2009) Surfatron acceleration in electromagnetic waves with a low phase velocity. *JETP Letters*, **89**, 441–447.

43 Vainchtein, D.L., Widloski, J., and Grigoriev, R.O. (2007) Mixing properties of steady flow in thermocapillary driven droplets. *Physics of Fluids*, 19:art, 067102.

44 Vainchtein, D.L., Widloski, J., and Grigoriev, R.O. (2007) Resonant chaotic mixing in a cellular flow. *Physical Review Letters*, 99:art, 094501.

45 Bogolyubov, N.N. and Mitropolsky, Yu.A. (1961) *Asymptotic Methods in the Theory of Nonlinear Oscillations*, vol. **537**, Gordon and Breach Science Publishers, New York.

46 Arnold, V.I. (1983) *Geometrical Methods in the Theory of Ordinary Differential Equations*, Springer, New York, Heidelberg, Berlin.

47 Timofeev, A.V. (1978) On the constancy of an adiabatic invariant when the nature of the motion changes. *Soviet Physics – JETP*, **48**, 656–659.

48 Neishtadt, A.I. (1987) On the change in the adiabatic invariant on crossing a separatrix in systems with two degrees of freedom. *PMM USSR*, **51**, 586–592.

49 Neishtadt, A.I., Vainshtein, D.L., and Vasiliev, A.A. (1998) Chaotic advection in a cubic Stokes flow. *Physica D*, **111**, 227–242.

50 Vainshtein, D.L., Vasiliev, A.A., and Neishtadt, A.I. (1996) Adiabatic chaos in a two-dimensional mapping. *Chaos*, **6**, 514–518.

51 Hughes, B.D. (1995) *Random Walks and Random Environments*, Oxford University Press, New York.

52 Dolgopyat, D. (2004) Repulsion from resonances. Preprint of U. Maryland.

53 Neishtadt, A.I., Simo, C., and Vasiliev, A.A. (2003) Geometric and statistical properties induced by separatrix crossings in volume-preserving systems. *Nonlinearity*, **16**, 521–557.

54 Grigoriev, R.O. and Schatz, M.F. (2006) Optically controlled mixing in microdroplets. *Lab Chip*, **6**, 1369.

55 Kroujiline, D. and Stone, H.A. (1999) Chaotic streamlines in steady bounded three-dimensional Stokes flows. *Physica D*, **130**, 105–132.

56 Neishtadt, A.I., Vainchtein, D.L., and Vasiliev, A. (2007) Adiabatic invariance in volume-preserving systems. in IUTAM Symposium on Hamiltonian Dynamics, Vortex Structures, Turbulence (eds. A.V. Borisov, V.V. Kozlov, I.S. Mamaev, and M.A. Sokolovskiy), pp. 89–108.

57 Solomon, T.H. and Mezic, I. (2003) Uniform, resonant chaotic mixing in fluid flows. *Nature*, **425**, 376–380.

58 Neishtadt, A.I. (2005) Capture into resonance and scattering on resonances in two-frequency systems. *Proceedings of the Steklov Institute of Mathematics*, **250**, 183–203.

59 Piro, O. and Feingold, M. (1988) Diffusion in three-dimensional Liouvillian maps. *Physical Review Letters*, **51**, 1709–1802.

60 Cartwright, J.H.E., Feingold, M., and Piro, O. (1995) Global diffusion in a realistic three-dimensional time-dependent nonturbulent fluid flow. *Physical Review Letters*, **75**, 3669–3672.

61 Cartwright, J.H.E., Feingold, M., and Piro, O. (1996) Chaotic advection in three-dimensional unsteady incompressible laminar flow. *Journal of Fluid Mechanics*, **316**, 259–284.

2
Fluid Stirring in a Tilted Rotating Tank
Thomas Ward

2.1
Introduction and Background Information

Developing processes to efficiently break up dispersed phase fluids is a topic of interest. This stems from recent interests in an overall reduction in the amount of energy required to provide certain materials that are derived by the processing of fluids. Examples of large and small-scale processes that utilize large amount of energy are the production of chemicals and emulsions using stirred tanks [1–3] and cements using tilted rotating tanks [4–7]. Consider a liquid filled-stirred tank, representing a bounded domain, in which an impeller is suspended in the fluid. Since the impeller is immersed in the liquid, the amount of energy required to rotate it is proportional to the fluid viscosity. This suggests that large viscosity fluids require more energy to stir at a given speed than a lower viscosity fluid and that increasing the rotation rate is costly even for low viscosity fluids, from an energy perspective, where it is a common industrial belief that higher rotation rates are required in order to produce efficient stirring. This may not always be the case because high rotation rates often lead to the appearance of KAM surface, which are barriers to stirring [8], that is, in KAM regions the mode of transport in a homogeneous fluid is diffusion limited. Breakup of these barriers has been extensively investigated [9–24] mostly in the context of liquid droplets representing the bounded domain where an initial volume, or area for an axisymmetric bounded flow [10, 13, 15, 17, 21–27], of fluid elements are stretched to exponential lengths over a finite amount of elapsed time by chaotic advection [28]. The result is mixing of the fluid in the bounded phase (drop) since the fluid element of exponentially increasing length fills the volume such that everywhere it is in close proximity to itself at a value less than the diffusion length $\hat{\ell}_D = \sqrt{\hat{D}\hat{t}}$ (where $\hat{D}$ is the molecular diffusivity, $\hat{t}$ is the elapsed time, and $\hat{\ }$ is used to denote dimensional variables). The practical application of this phenomenon is a translating drop where rapid exchange of material between the inner and outer phases is desirable, a process which is often limited to slow diffusion [29] even at speeds where convection is relevant to momentum transport.

Transport and Mixing in Laminar Flows: From Microfluidics to Oceanic Currents,
First Edition. Edited by Roman Grigoriev.
© 2012 Wiley-VCH Verlag GmbH & Co. KGaA. Published 2012 by Wiley-VCH Verlag GmbH & Co. KGaA.

Overcoming slow diffusion in KAM regions in stirred tanks typically involves developing a stirring technique tailored to a specific system of interest. This may involve placing baffles in a stirred tank [3] or multiple impellers [2], or adding a herringbone pattern to a channel wall in a microfluidic device [30] and other geometric features to help stirring. The one feature that most of these techniques have in common is the generation of some form of flow vorticity. More precisely, the goal is often to generate vorticity that is spatially and temporally inhomogeneous. Vorticity is associated with flow rotation since its magnitude and direction are calculated by taking the curl of the velocity. So the real question in developing stirring protocols is how to control/enhance rotation? Before one can control rotation in their specific flow it may be worthwhile to examine why rotation is important in the stirring, and eventual mixing, of fluids. Then one can design these attributes into their particular flow, be it a fluid in a microchannel or a large industrial-sized system for stirring.

Stretching has been identified as one of the components (the others are folding and twisting) necessary for stirring of fluid element that ultimately leads to mixing in the case of a homogeneous fluid. Periodic shearing [31–33] involves the study of generating efficient stirring protocols by using linear velocity distributions in a periodic, usually planar Cartesian coordinate, domain. Since the domain is periodic, an initial fluid element will eventually come into close proximity with itself after a finite amount of time through stretching. The periodic shear stretching process, in a finite domain, is performed over a sufficient amount of elapsed time until the distance between adjacent portion of the fluid element is less than the diffusion length, and mixing has occurred, similar to the process in a bounded domain inside of a drop. The temporal inhomogeneity in shear is analogous to gradients (spatial inhomogeneity) and lead to optimal mixing under certain conditions [32].

Stretching in a rotating geometry such as the Taylor-Couette cell is the polar coordinate analog of the periodic shear flow problem in Cartesian coordinates. Stretching in a Taylor-Couette cell is unique due to the fact that after one revolution the fluid begins to resemble an Archimedean spiral (periodic boundaries) under certain conditions that are related to the distribution of voricity in the flow. To continue this discussion on fluid element stretching in the context of vorticity consider the flow that develops due to a fluid rotating between concentric cylinders (Taylor-Couette) and the solution proposed by Taylor [34] in the limit of small gap spacing (the solution for any gap spacing was described in more recent years [35]). Let the outer cylinder have radius $\hat{r}_1$ and rotate with the angular rotation rate $\hat{\Omega}_1$ and let the inner cylinder have radius $\hat{r}_2$ and rotate with the angular rotation rate $\hat{\Omega}_2$. Taylor's general solution is for the flow in the gap when the spacing is small, that is, $(\hat{r}_1 - \hat{r}_2)/\hat{r}_1 \ll 1$, and yield the equations of motion $\hat{v}_\theta(\hat{r}) = \hat{C}_1 \hat{r} + \frac{\hat{C}_2}{\hat{r}}$ and $\hat{v}_r = 0$ for the azimuthal and radial velocity components, respectively (the flow is consider homogeneous in the z direction). The constants are $\hat{C}_1 = \hat{\Omega}_1 - \hat{C}_3(\hat{r}_1, \hat{r}_2) \left(\frac{\hat{\Omega}_2 - \hat{\Omega}_1}{1 - \hat{C}_3(\hat{r}_1, \hat{r}_2)} \right)$ and $\hat{C}_2 = \frac{\hat{r}_2^2 (\hat{\Omega}_2 - \hat{\Omega}_1)}{1 - \hat{C}_3(\hat{r}_1, \hat{r}_2)}$, where $\hat{C}_3(\hat{r}_1, \hat{r}_2) = \frac{\hat{r}_2^2}{\hat{r}_1^2}$. Inserting the velocity into the advection equation $\hat{r}\dot{\theta} = \hat{v}_\theta$ and one integration yields the relationship $\theta(\hat{r}) = \hat{\omega}(\hat{r})t$, where

2.1 Introduction and Background Information

$\hat{\omega}(\hat{r}) = \hat{C}_1 + \frac{\hat{C}_2}{\hat{r}^2}$. $\theta(\hat{r})$ is the orbit representation of the flow between the concentric rotating cylinders. The orbits are closed trajectories that lie in a plane, and particles that lie along the orbits move with an angular speed $\hat{\omega}(\hat{r})$. The deformation of fluid elements due to the flow is given by the difference of adjacent particles, that is, displaced by passively tracing a given set of orbits as a function of time. The maximum deformation occurs for a fluid element that spans the gap and the amount of deformation is given by the difference $\Delta\theta(\hat{r}) = \theta(\hat{r}_2) - \theta(\hat{r}_1) = \hat{C}_2 \left(\frac{1}{\hat{r}_2^2} - \frac{1}{\hat{r}_1^2} \right) \hat{t}$. $\hat{C}_2 = 0$ for values of $\hat{\Omega}_1 = \hat{\Omega}_2$ or $\hat{r}_2 = 0$ suggesting that $\Delta\theta = 0$. In either of these situations, there are no radial gradients in the velocity; therefore, all the orbits have a speed that varies linearly in the radial direction. This is the definition for a rotational vortex which is characterized by an absence of stretching of fluid elements, that is, the length of an initial material line spanning the gap $[\hat{r}_1, \hat{r}_2]$ remains constant and executes solid body rotation. For values of $\hat{\Omega}_1 \neq \hat{\Omega}_2$ or $\hat{r}_2 \neq 0$ the orbit rate varies nonlinearly in the radial direction and the vorticity is spatially inhomogeneous. The consequences of this are stretching of fluid elements spanning the gap and that the length $\Delta\theta \neq 0$ is a function of time. While the process eventually leads to mixing by stretching of fluid elements, Lyapunov exponents are essentially zero since fluid elements separate only linearly with time, instead of exponential [36].

In this chapter, it will be shown that the flow of fluid in a *tilted-rotating tank* leads to gradients in orbits (rotational speed) in the vertical direction of the form

$$\theta(\hat{z}) = \hat{\omega}(\hat{z})\hat{t} \qquad (2.1)$$

in the vicinity of the free surface. The angular velocity in the vicinity of the free surface is always less than or equal to the angular velocity of the tank itself whenever a fluid filled tank is rotated in a gravitational field. The gradients lead to the appearance of a flow that resembles periodic shearing. The gradients alone does not lead to efficient stirring but are essential to the stirring process since the gradients always lead to stretching of fluid elements resulting in a process that puts them into closer proximity in a finite amount of time.

At the free surface (air–liquid interface) the bulk fluid cannot accurately track the rotating tank motion (solid-body rotation) since gravity will tend to level the free surface. This guarantees a break in the symmetry in the flow since the local tank velocity is always asymmetric on opposite sides spanning the free surface, for example consider the portion of the tank where the liquid is drawn along the tank walls in a direction that is opposite to gravity. It leads to a coating [37–42] and drainage flow of the fluid along the tank walls under certain wetting conditions. Just below the surface, in the regions where the fluid wets the walls and drains, there will be a small vortical flow that develops [39]. The small vortical flow in the vicinity of the free surface is very important to the stirring process in the tilted-rotating tank since these vortices occur in the azimuthal direction in a polar geometry generating a torodial wall vortex and can therefore facilitate the vertical and radial transport of fluid. Although the vertical and radial transport is desirable, there is no stretching of fluid elements and ultimately any efficient mixing due to the vortical flow is driven by pure diffusion, that is, the vortical flow due to the fluid drainage along the wall generates KAM surfaces. But these can be broken up by periodic shearing assuming the vortices transport material into the vicinity of the liquid's rotation axis.

The convective transport of a nonneutrally buoyant liquid-phase material will be investigated in the laminar flow limit of the tilted rotating tank. Successful stirring would be particularly useful for the production of emulsion. The fact that the flow is mostly shear driven, stretching is also beneficial to stirring of liquid–liquid phases since this would facilitate breakup of the dispersed phase fluid [43–45]. For liquid–liquid or solid–liquid stirring, the possible convective transport of denser material to the vicinity of free surface and back is also desirable.

Numerical and experimental Poincaré mapping will be used to understand the periodic shearing process in the tilted rotating tank geometry [46, 47]. The numerical Poincaré mapping consists of three-dimensional finite time Poincaré mapping to understand the dispersive quality of the tilted rotating tank flow. In this mapping, many particle trajectories are sampled with a constant frequency [19]. Experimental Poincaré mapping will be used to qualitatively understand flow characteristics for comparison with the analysis presented in the following section. Three experimental setups are considered: (1) low Reynolds number homogeneous flow, (2) $O(1)$ Reynolds number homogeneous flow, and (3) $O(10)$ Reynolds number inhomogeneous flow.

In the next section, the analysis of a titled rotating tank is presented by performing a power series expansion of the solid-body rotation in a tilted geometry. This is followed by a presentation of the asymptotic limits for the flow and of the analytical results. Then the experiments are described along with the results. Finally, there is some discussion of the experiment and theory along with some concluding remarks.

2.2
Tilted-Rotating Tank Analysis

2.2.1
Tilted-Rotating Tank Model Equation

Consider a Newtonian fluid of volume $\hat{V}$, with viscosity $\hat{\mu}$, and density $\hat{\varrho}$, rotating in a titled tank. The tank rotates with the rotation rate denoted by $\hat{\Omega}_0$, and the tank radius $\hat{R}$, about the rigid body rotation axis z_a and is tilted with an angle, α (relative to the horizontal), about the y-axis (see Figure 2.1). Cartesian coordinates are used to describe the system where all spatial variables are made dimensionless using $\hat{R}$, time using $\hat{\Omega}_0$, velocity using $\hat{R}\hat{\Omega}_0$, and stress using $\hat{\Omega}_0\hat{\mu}$. The dimensionless free surface, located at $z = 1/\lambda$, where $\lambda = \hat{R}/\hat{H}_0$, remains flat during rotation provided the capillary number $Ca = \hat{\mu}\hat{\Omega}_0\hat{R}\hat{\gamma}$, where $\hat{\gamma}$ being the surface tension, is less than unity. For a partially filled tank in the limit of negligible convective momentum transport, or equivalently small Reynolds number, $Re = \hat{\varrho}\hat{\Omega}_0\hat{R}^2\hat{\mu} \ll 1$, the slow rotation causes a disturbance to the flow in the vicinity of the free surface. The dimensionless Stokes

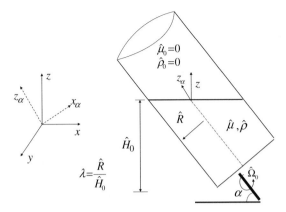

Figure 2.1 Schematic of titled-rotating tank system. The tank of radius $\hat{R}$ is tilted with angle α relative to the horizontal and rotates with speed $\hat{\Omega}_0$. The distance from the lowest vertex in elevation of fluid in the tank to the free surface is the liquid height denoted as $\hat{H}_0$. The Cartesian coordinate system is chosen such that the tank is rotated at the y-axis.

flow equations governing momentum transport and continuity for the rotating-tilted tank are $\nabla \cdot \sigma = 0$ where $\sigma = -P\mathbf{I} + \tau$ and $\nabla \cdot \mathbf{v} = 0$, respectively. $\tau = \mu[\nabla \mathbf{v} + \nabla \mathbf{v}^T]$ is the viscous stress tensor.

The flow in a tilted rotating tank is generated by a combination of a two-dimensional rigid-body rotation in the xy-plane and in the yz-plane. A disturbance to the linear equations in the titled frame is now considered through boundary conditions for the free surface and through satisfying continuity. These terms will appear as the leading-order nonlinear terms. In order for the proposed analytical technique to be valid, there must be a region in the fluid domain where there is a principal axis about which the fluid rotates. Therefore, it is assumed that the total flow, which is a combination of the rigid-body rotation in the two planes, still possess an axis about which the fluid rotates called a liquid rotation axis. The liquid rotation axis is defined as the location where $\mathbf{v}(\tilde{\mathbf{x}}) = 0$ (˜ denotes variables used to represent location along the liquid rotation axis throughout this chapter). The properties of the liquid rotation axis, $\tilde{\mathbf{x}}$, are important in determining local and global aspects of the composite flow. For now, the assumption of a liquid rotation axis allows one to expand the linear tilted-rotating flow in the vicinity of the zero velocity region by adding nonlinear terms representing the disturbed flow portion of the equations of motion. Then the boundary conditions may be satisfied in the vicinity of the free surface.

The general form for the perturbed autonomous advection equation $\dot{\mathbf{x}} = \mathbf{v}(\mathbf{x})$ representing the equation of motion governing the proposed composite flow of the rigid-body rotation of a liquid filled tank perturbed by a tilted geometry is

$$\begin{pmatrix} \dot{x} \\ \dot{y} \\ \dot{z} \end{pmatrix} = \begin{pmatrix} y\cos\alpha \\ 1-x\cos\alpha - z\sin\alpha \\ y\sin\alpha \end{pmatrix} + \begin{pmatrix} 0 \\ (f(y)+g(z))\sin\alpha \\ h(y,z)\sin\alpha \end{pmatrix} \qquad (2.2)$$

where the first term is the linear flow that results from the superposition of two rigid-body rotations and the last vector contains the leading-order terms to the disturbed flow. The disturbed flow portion represents the minimum number of terms required to satisfy the boundary conditions discussed in the following.

The stress components must satisfy $e_z \cdot \tau \cdot e_x = 0$ and $e_z \cdot \tau \cdot e_y = 0$ along the free surface at $z = 1/\lambda$. The vertical velocity component vanishes along the free surface, or $e_z \cdot v = 0$, and the fluid must be incompressible, or $\nabla \cdot v = 0$. A power series for each unknown term is assumed for each of the unknown functions, that is, $f(y) = C_1 y^p$, $g(z) = C_2 z^q$, and $h(y,z) = C_3 y z^r$. The function $h(y,z)$ must be linear in y to satisfy the zero normal velocity boundary condition along the free surface. After applying the boundary conditions, $\frac{\partial u}{\partial z} = 0$, $\frac{\partial v}{\partial z} = 0$, and $w = 0$ at $z = 1/\lambda$, and continuity, the equations of motion near the liquid rotation axis for a rotating partially filled tilted tank are

$$\begin{pmatrix} \dot{x} \\ \dot{y} \\ \dot{z} \end{pmatrix} = \begin{pmatrix} y \cos \alpha \\ 1 - x \cos \alpha - z \sin \alpha + \frac{\lambda}{2}(z^2 + y^2) \sin \alpha \\ y \sin \alpha (1 - \lambda z) \end{pmatrix} \quad (2.3)$$

The principal axis is equal to the liquid rotation axis that lies in the xz-plane, $\tilde{x} = (\tilde{x}(\tilde{z}), 0, \tilde{z})$ where

$$\tilde{x}(\tilde{z}, \alpha) = \left(\frac{\lambda}{2}\tilde{z}^2 - \tilde{z}\right) \tan \alpha + \sec \alpha. \quad (2.4)$$

This quadratic equation, of constant curvature $\frac{d^2 \tilde{x}}{d\tilde{z}^2} = \lambda \tan \alpha$, is a parabola representing a family of fixed points emanating from the free surface at $1/\lambda$ and terminate at the rotating tank axis where the fluid rotates with the same speed as the tank.

2.2.1.1 Asymptotic Analysis: Free Surface Vortex

The first analysis of the equations of motion for the tilted-rotating tank consists of considering an asymptotic limit. The goal is to determine approximate solutions for the equations in the vicinity of the free surface and near the rotating tank walls in the $x_\alpha y$-plane.

The dimensionless inverse distance to the free surface λ and the tilt angle α are two parameters that will be simultaneously examined in their asymptotic limits. Physical interpretation of the asymptotically small limits for these two parameters is similar, that is, a small tilt angle suggests that the region of disturbed fluid flow is simultaneously small, while a large relative distance to the free surface suggests that most of the fluid domain consists of undisturbed fluid that rotates with the same speed as that of the tank. In the limit of a vanishingly small tilt angle α or extremely tall tanks where $\lambda \to 0$ the angle is chosen such that $\sin \alpha \sim \varepsilon$, $\cos \alpha \sim 1$, and tank height such that $\lambda \sim \varepsilon$. Inserting these expressions into the tilted-rotating tank model equation and ignoring $O(\varepsilon^2)$ terms yield the resulting linear equation

$$\dot{x}_\varepsilon = (y, 1-x, 0) + \varepsilon(0, -z, y) \tag{2.5}$$

where the two terms denote the unperturbed and perturbed flows, respectively.

The general solution of the reduced equation is

$$x_\varepsilon(t) = (1 + x_\varepsilon, y_\varepsilon, z_\varepsilon) = (1 + \cos f(\varepsilon)t, -f(\varepsilon)\sin f(\varepsilon)t, \varepsilon \cos f(\varepsilon)t) \tag{2.6}$$

where $f(\varepsilon) = \sqrt{\varepsilon^2 + 1}$. The trajectories $x_\varepsilon(t)$ are closed orbits since $|x_\varepsilon| = 1 + \varepsilon^2$, that is, independent of time. The fluid speed in the vertical direction and near the tank walls is approximately equal to $|dz_\varepsilon/dt| = \varepsilon$, and is the approximate angular speed (azimuthal frame of reference) of the torodial wall vortex that form beneath the free surface (see Figure 2.2). Assuming that the wall vortex region, beneath the surface, is of vertical and horizontal size ε then the approximate cross-sectional area of stirred fluid is the triangle of height $O(\varepsilon)$ (measured vertically downward from the free surface) and the triangle base measured inward from the tank wall.

The liquid rotation axis in the asymptotic limit is

$$\tilde{x}(\tilde{z}, \varepsilon) = 1 - \varepsilon \tilde{z} \tag{2.7}$$

or a line with slope of magnitude $-\varepsilon$ where higher order terms, $O(\varepsilon^2)$, are clearly responsible for the curvature in the liquid rotation axis. Given that the tilt angle is small such that $\sin \alpha \sim \varepsilon$, the liquid rotation axis and tank walls are parallel in this asymptotic limit. Also, given that the wall vortex cannot occupy both sides of the cross-sectional area domain, its maximum depth is $\sin \alpha = \lambda = 0.25$. The $O(\varepsilon)$ curvature term in the liquid rotation axis is not negligible for the values of $\sin \alpha$ larger than the $O(\varepsilon)$ limit and an analysis in the vicinity of the axis instead of the analysis in the vicinity of the tank walls performed here is more appropriate for determining the stirring characteristic of the tilted rotating tank [48].

This approximate form to the equations of motion still satisfies continuity and the zero shear stress boundary condition along the free surface but not the zero vertical

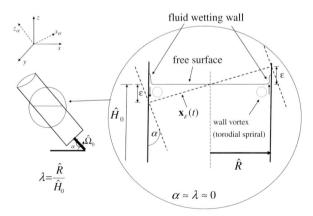

Figure 2.2 Illustration of uneven free surface in the tilted-rotating tank system in the asymptotic limit of the small tilt angle, $\sin \alpha \sim \varepsilon$, and/or very large liquid heights, $\lambda = \hat{R}/\hat{H}_0 \sim \varepsilon$.

velocity component. Therefore, the no slip condition may be violated allowing fluid slip along the wall. This motion will tend to drive fluid upward along the wall when y is positive and downward when it is negative for all tilt angle values greater than zero.

2.2.1.2 Linear Analysis: Periodic Shearing

The angular velocity $\Omega = \frac{1}{2} \nabla \times \dot{x}$ for the approximate equations of motion has the form

$$\Omega = (\Omega_x, 0, \Omega_z) = ((\lambda z - 1)\sin \alpha, 0, -\cos \alpha) \qquad (2.8)$$

According to this expression the system has two principal axis for the angular velocity e_x and e_z. The Ω_z component is constant for a fixed value of the tilt angle α but the Ω_x component varies linearly in the z-direction such that its value is zero at the free surface. This is a direct result of the no penetration boundary condition at the free surface.

The approximate equations of motion for the flow in a tilted rotating tank were formed by expanding the linear equations for solid body rotation using the minimal boundary conditions needed to produce a nonlinear flow. The assumptions and approximations lead to the realization that a liquid rotation axis, $\tilde{x}$ (Eq. (2.4)), exists for this system where $v(\tilde{x}) = 0$. An expansion about this principal axis is now performed in order to show the existence of vertical angular velocity gradients in its vicinity. This analysis is similar to one performed previously [48] and the reader should refer to that manuscript for a more complete analysis.

The tangent line to the liquid rotation axis is

$$\tilde{m}(\tilde{z}) = (\lambda \tilde{z} - 1) \tan \alpha \qquad (2.9)$$

with the normalized vector tangent

$$\hat{m} = (m_x, m_y, m_z) = \frac{1}{\sqrt{1 + \tilde{m}(\tilde{z})^2}} (\tilde{m}(\tilde{z}), 0, 1). \qquad (2.10)$$

The nonlinear equations of motion are linearized about the rotation axis x using the eigenvalue equation $[A - I\omega(\tilde{z})]\mathbf{k} = 0$ where $A = (\nabla v)^T$, and I is the identity tensor. One eigenvalue is zero, so that the solution is constant and the associated eigenvector is set equal to zero. The two other eigenvalues are always purely imaginary and lead to a solution for the linear equation of the form

$$\mathbf{x}(\theta(\tilde{z})) = (x, y, z) = \left(\frac{\cos \alpha}{\omega(\tilde{z})} \sin \theta(\tilde{z}), \cos \theta(\tilde{z}), -\frac{\tilde{m}(\tilde{z}) \cos \alpha}{\omega(\tilde{z})} \sin \theta(\tilde{z}) \right) \qquad (2.11)$$

where $\theta(\tilde{z}) = \omega(\tilde{z}) t$ is the phase angle with the orbit frequency

$$\omega(\tilde{z}) = \cos \alpha \sqrt{1 + \tilde{m}(\tilde{z})^2} \qquad (2.12)$$

Since the eigenvalues were determined by linearizing the nonlinear model and, have no real parts, the stability of the system of equations is indeterminable.

Using Eqs. (10) and (11) we see that $\mathbf{x} \cdot \tilde{\mathbf{m}} = 0$, so that the trajectory planes are perpendicular to the liquid rotation axis. The trajectories, $\mathbf{x}(\theta(\tilde{z}))$, are a family of two-dimensional circular orbits of the constant radius, $|x(\theta(\tilde{z}))| = 1$, similar to solid body rotation. The difference between this flow and solid-body rotation is that the orientation of the streamlines and speed depend on their vertical location z. We see from Eq. (2.12) that the critical point location where $\omega(\tilde{z}_{crit}) = 1$ is at $\tilde{z}_{crit} = 0$.

The horizontal distance between the intersection of the liquid rotation and solid-body rotation axis, $\tilde{x}(\tilde{z}_{crit} = 0, \alpha)$ (see Figure 2.3), and the liquid rotation axis at the free surface, $\tilde{x}(\tilde{z} = 1/\lambda_{crit}(\alpha) = 2 \sin \alpha)$, diverges as

$$\Delta \tilde{x} = \sin \alpha \tan \alpha \qquad (2.13)$$

So while the vertical height of the liquid rotation axis is bounded, $0 < 1/\lambda_{crit}(\alpha) < 2$ the horizontal dimension of the rotation axis diverges like $\tan \alpha$ for $0° \ \alpha \ 90°$. The angular velocity gradient is $\propto 1 - \cos \alpha$ and increases with an increase in the value of the tilt angle and so does the cross-sectional area (as measured by the region spanning the liquid rotation axis) it covers. In the limit $\Delta \tilde{x} > 1/\lambda$ the stirring dynamics are dominated by periodic shearing. This geometric interpretation of efficient stirring in a tilted rotating tank, that is, maximizing the angular velocity gradient in the cross-sectional area region spanning the free surface,

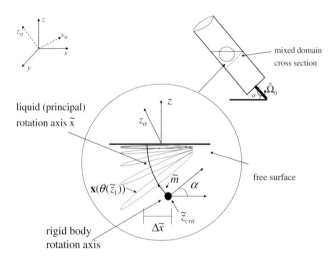

Figure 2.3 Illustration of flow in the vicinity of the liquid rotation axis. The approximate flow in this region is a set of orbits $\theta(z)$ with speed variations in the vertical direction due to the curvature of the liquid rotation axis. The horizontal length, $\tilde{x}$, of the liquid rotation axis varies with the tilt angle α and is independent of the liquid volume in the tank for volumes of fluid $\hat{V} > \hat{R}^3 \tan \alpha$.

led to an approximate optimal mixing tilt angle of $\alpha \approx 53°$ [48] for the low Reynolds number stirring in a tilted rotating tank for certain fluid volumes.

2.2.2
Comments on Laminar Flow

It has already been observed that the low Reynolds number case always yields a torodial wall vortex and the structure is independent of the volume of fluid as long as the value for the vertical location of the fluid interface is larger than the value for the bottom wall, or equivalently the condition $= \hat{V}/(\pi \hat{R}^3) \geq \tan \alpha$ (derived from geometry) is satisfied. An angular velocity gradient in the vertical direction is generated because of the curvature in the liquid rotation axis. This velocity gradient is small in the low Reynolds number case and generates a self-similar wall vortex structure. The fluid in the vortex region may combine with the periodic shearing near the liquid rotation axis in a process that qualitatively resembles chaotic advection [28, 33]. However, previously it was observed that the overall rate of mixing did not occur at an exponential rate of mixing, similar to that typically associated with chaotic advection at low Reynolds numbers. Instead, a transition from an initially linear to a logarithmic profile was observed [48].

For larger values of the Reynolds number, $O(> 1)$, where laminar type flow is expected, the situation changes in the tilted rotating tank. Viscous forces, associated with low Reynolds numbers, are dominated by the flow geometry, that is so it is not surprising that the low Reynolds number flow in a tilted rotating tank produced self-similar structures that are independent of fluid volume above certain values [48]. Inertial forces associated with high values of the Reynolds number result in large characteristic speeds of the fluid, and often negligible wall effects. Laminar flow is characterized by a balance between the inertial and the viscous forces.

In the laminar flow regime, the angular velocity gradient increases in strength. For the tilted rotating tank geometry, this will result in an interaction between the torodial wall vortex and the cylindrical tanks bottom wall due to viscous dissipation, generating additional secondary vortices in a cascade that is reminiscent of the Moffatt vortices [49]. The secondary vortices do not stir fluid like the torodial wall vortex because the wall vortex still interacts with trajectories in the vicinity $x \approx \tilde{x}$ so periodic shearing (and the subsequent breakup of these regions) occurs, while the secondary vortices are generated by a combination of inertia and viscous dissipation due to the presence of the bottom wall. However, the secondary vortices may help in stirring by convective transport of material from the bottom wall. This is especially useful for mixing emulsions.

However, large inertia does not always guarantee good mixing. Without velocity gradients, there is little transport of material from the walls to the bulk fluid. This can be especially problematic in very large rotating tank systems that inherently operate at large Reynolds numbers since its value is proportional to the tank radius squared. Indeed this is what has been observed in some rotating tank systems where the stirring in large tanks at high Reynolds numbers is confined to regions close to walls [1].

2.2.3
Analytical Results

In this section, the analytical results are summarized in Figure 2.4. Solutions to the nonlinear equations of motion, Eq. (2.3), are plotted in Figure 2.5 and are found by using an implicit fourth-order Runge–Kutta algorithm with explicit starter [23, 50]. The equations are integrated to machine precision which is needed to ensure particles that start on one trajectory do not drift to other nearby trajectories and give false results of volume filling solutions. Even though this high order of accuracy is imposed there is still some drift leading to the conclusion that the solutions may be volume filling (dispersive) when integrated over a large enough timescale.

Figure 2.4a are plots of the asymptotic flow in the vicinity of the free surface (Eq. (2.6)) for three values of ε. The xz- and yz-plane views are shown for three values of the small parameter $\varepsilon = 0.1$, 0.2, and 0.3. The amount of tilt is proportional to the value of ε. Gravity will level this surface and the process leads to the appearance of a torodial wall vortex. The magnitude is proportional to the value of ε. For a larger value of ε, there is appreciable curvature of the liquid rotation axis. In the vicinity of the liquid rotation axis the trajectories are given by the solution to the linearized equations, Eq. (2.9). In this region there is periodic shearing, which is the proposed mechanism that generates stretching in the tilted rotating tank.

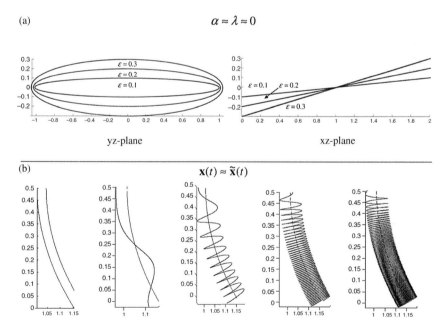

Figure 2.4 Results from analytical solution presented for (a) the asymptotic analysis in the vicinity of the free surface and (b) the flow in the vicinity of the liquid rotation axis for $\alpha = 1/\lambda = \pi/6$.

Figure 2.4b are plots of an initial material line that is used to illustrate the periodic shearing process. These plots are generated with the initial material line lying in the same plane, $y = 0$, as the rotation axis. The time interval is chosen to be near the natural frequency of the tank. As time advances, so does the angular displacement between nearby trajectories. The material line can never lie in the initial plane for time $t > 0$ because the rate of deformation, $\dot{\Delta\theta}$, varies in the vertical direction since $\omega \propto z$ for all $\alpha > 0$. The material line begins to fill a tubular surface that is a plane wrapped around the rotation axis as time approaches infinity represented by the last image in the Figure 2.4b set.

Poincaré map results are shown in Figure 2.5. There are 2500 initial conditions in close proximity shown in Figure 2.5a ($\mathbf{x}_0 = (1, y_0, z_0)$ where $y_0, z_0 \in [0.3, 0.4]$ in increments of 0.002) in a rectangular shape representing an initial fluid element. After approximately 10 periods (there are approximately 7.2 dimensionless units of time in one period) dispersion of the particles can be seen through stretching of the initial rectangular fluid element. In Figure 2.5b–f, the process is continued until the initially rectangular patch of particles has deformed into a spiral shape. There is both vertical and radial dispersion of the particles from their initial rectangular arrangement that can be seen if Figure 2.5f.

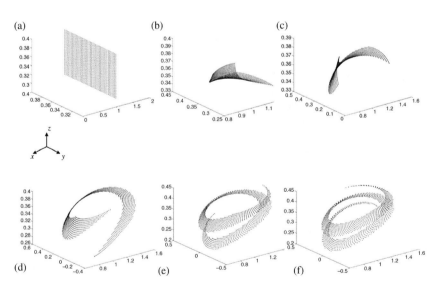

Figure 2.5 Numerical solutions to Eq. (2.3) for 2500 initial conditions for $\alpha = 1/\lambda = 30°$. Each image frame shows a snapshot in time of the location of the particles for a fixed frequency or a three-dimensional Poincaré map. The images are produced after (a) 0, (b) 10, (c) 20, (d) 50, (e) 80, and (f) 100 periods.

2.3 Experiments

2.3.1 Setup and Procedure

2.3.1.1 Low Reynolds Number Experiments: Homogeneous Fluid

Low Reynolds number experiments were performed using a cylindrical acrylic tank 3 inches in diameter and 48 inches in length. The tank was mounted to a large piece of acrylic using plastic dampers. A solid cylindrical piece was used as an end cap to mount the cylinder to the motor. The end cap has a large hole drilled in the middle with a solid metal bar glued in this slot. The motor was mounted to a piece of acrylic with a hole for the drive shaft that attaches to the metal bar with a custom fabricated coupling. For the long tube it is essential to adjust the coupling prior to each experiment to ensure that the shaft is parallel with respect to the tube. The motor is tested and set to rotate with a constant frequency of approximately 0.1 Hz. At the top of the acrylic mounting piece was a mounting plate with a small hole drilled for the laser. A planar laser sheet was generated using a lasiris laser (StockerYale) which emits a green sheet (wavelength 533.1 nm) at a constant output power of 5 mW.

The fluid used was pure glycerol (Clearco) which has a viscosity of approximately 0.93 Pa s and a density of 1260 kg m^{-3} at room temp 20 °C both numbers taken from tabulated values. A small portion of the glycerol was dyed with laser grade Rhodamine B (Sigma–Aldrich) until the solution became uniform and transparent but fluorescent, in color. The dye solution was kept from ambient light that causes the dye to degrade and the luminescence to weaken in intensity. Based on our physical parameter, dimensions, and motor speed the Reynolds number was estimated to be approximately $Re = 0.1$ for this set of experiments, with $0.3 < \lambda < 0.75$ and $0.09 < \sin \alpha < 0.91$ (the range of angles is $5° < \alpha < 65°$ in $10°$ increments for this set of experiments). The capillary number is approximately, $Ca = 0.07$ with an air-glycerol-acrylic surface tension of approximately 55 mN/m taken from tabulated values.

The experimental Poincaré maps were generated by, first, filling the tank with a known volume of fluid. Then the distance from the free surface to the minimum fluid location in the bottom of the tank was measured and recorded. A small blob of the dye solution was placed near the free surface and the laser was placed in the mounting plate. The laser sheet was placed so that it is illuminated the $y = 0$ plane that contains the liquid rotation axis. The motor was turned on and the flow plane perpendicular to the laser sheet was viewed from the side using a CCD camera (Sony X710). There was some optical distortion since the glycerol and acrylic were not indexed match, but the indices of refraction are similar and the optical distortion was minimal. Images were taken in set intervals of 10 s for approximately 10^4 s, so each experimental run contains nearly 10^3 frames.

2.3.1.2 Laminar Flow Experiments: Homogeneous and Inhomogeneous Fluids

O(1) Reynolds Number Experiments: Homogeneous Fluid $O(1)$ Reynolds number experiments were performed using a cylindrical acrylic tank of approximately 3 inches in diameter and 12 inches in length. The tank was mounted on a large piece of acrylic using plastic dampers to ensure smooth operation. A solid cylindrical piece was used at the bottom as an end cap, and also to mount the cylinder to the motor. The end cap has a large hole drilled in the middle with a solid metal bar attached to the slot with acrylic glue. The motor was mounted to a piece of acrylic with a hole for the drive shaft that attaches to the metal bar with a custom-fabricated coupling. The motor was tested and set to rotate with a constant frequency of approximately 0.3, 0.4, and 0.8 Hz depending on the input voltages of either 4.5, 6, or 12 V. At the top of the acrylic mounting piece was a mounting plate with a small hole drilled for the laser. A planar laser sheet was generated using a lasiris laser (StockerYale) which emits green sheet (wavelength 533.1 nm) at a constant output power of 5 mW. The planar laser sheet was positioned to illuminate the central bisecting plane of the tank ($y = 0$ plane).

The fluid was a mixture of 90% pure glycerol (Clearco) and 10% distilled water which has a combined kinematic viscosity of approximately 250 cSt estimated from tabulated values [51, 52]. A small portion of the glycerol/water mixture was dyed with laser grade Rhodamine B (Sigma–Aldrich) until the solution became uniform and transparent but fluorescent, color. Based on our physical parameter, dimensions, and motor speed we estimate the Reynolds number to be in the range $1.1 < Re < 5.1$ in these experiments. The viscosity of glycerol varies with small changes in temperature and water concentration so the range of Reynolds numbers are approximate values.

The experiments were performed by, first, filling the tank with a known volume of fluid either $\hat{V} = 200$ or 400 mL. Based on these fluid volumes and tilt angle range we expect the fluid interface to clear the bottom wall, that is, $\hat{V} \geq \hat{R}^3 \tan \alpha$ for the 35° angle experiments shown. At this tilt angle the values of λ are 0.75 and 0.30 for the 200 and 400 mL experiments, respectively.

A small blob of the dye solution was placed near the free surface and the laser was placed in the mounting plate. The laser sheet was placed so that it illuminated the $y = 0$ plane (central bisecting plane). The motor was powered on and the flow plane perpendicular to the laser sheet was viewed from the side using a CCD camera (Sony X710). There was some optical distortion since the glycerol mixture and acrylic are not precisely indexed match, but the indices of refraction are very similar (more so than pure glycerol and acrylic) and the optical distortion was minimal in all of the experiments reported. Images were taken in set time intervals of 0.5 to 10 s, depending on rotation speed where larger intervals are used for slower rotations. Images are taken over a total elapsed time spanning approximately 10^4 s, so each experimental run contains at least 10^3 frames based on the maximum amount for a time interval in an experiment of 10 s. The experiments do not reach an absolute steady state because diffusion (slow mixing process) can possibly lead to the appearance of further stirring if the domain has not completely mixed by the end of an experiment.

2.3 Experiments

O(10) Reynolds Number Experiments: Inhomogeneous Fluid The physical setup for the inhomogeneous fluid experiments were identical to the laminar homogeneous experiments described previously. The only difference is the working fluids: the continuous phase fluid is a light mineral oil with approximate viscosity of 30 cSt and density of approximately 890 kg/m^3. This reduced viscosity means the Reynolds numbers were larger than the value in the inhomogeneous fluid experiment by a factor of about 10, so that the Reynolds numbers are $O(10)$. The capillary number will also decrease due to the decreased viscosity with $Ca \sim O(0.001)$. The dispersed phase fluid was deionized water with a small amount of surfactant (Palmolive) added which is needed to help facilitate breakup of the large drop. The concentration of surfactant was either $\phi = 0.01$, 0.02, 0.04, and 0.1 by volume. The surfactant concentrations were low such that the physical properties are assumed to be that of pure water with a viscosity of 1 cSt and density 1000 kg m^{-3}.

The experimental procedure consisted of injecting 5 mL of the water-surfactant solution into the tank filled with 200 mL of light mineral oil, prior to initiating rotation. The solution is allowed to settle into the bottom of the tank and then the motor is turned on. The index of refraction between the water and oil is low such that the large 5 mL drop initially appears invisible to the laser light. The large drop begins to breakup into smaller droplets after a sufficient amount of elapsed time. When the droplets reach a diameter that is less than the width of the laser sheet (approximately 800 μm in width) then the drops begin to internally refract the laser light. After longer time, there is more breakup of larger drops into smaller ones and a higher intensity of the refracted laser light. Therefore, the light intensity is proportional to the number of drops that are present with a diameter less than the laser sheet width and relatively higher light intensity, when comparing one experiment to another, suggest more droplets with diameter smaller than the laser sheet, and hence more efficient stirring.

2.3.2 Results and Analysis

2.3.2.1 Low Reynolds Number

Figures 2.6–2.8 show the initial and final condition for experiments performed at low Reynolds number, $Re = 0.1$ with tilt angles ranging from $\alpha = 5°$ to $65°$ in $10°$ increments (the $\sin \alpha$ of the angles are shown in the figures). The dimensionless heights, $1/\lambda$, are shown in the left-hand column for this series of images.

In Figure 2.6a the experiments are performed at the smallest tilt angle of $\alpha = 5°$. There is little difference between the initial and final conditions except for the appearance of small vortex structures near the tank wall on the either side. The depth below the surface of the vortices appears to be on the order of $\hat{R} \sin \alpha$ as predicted for the small tilt angle and small λ limit where the asymptotically small parameter value is $\lambda \sin \alpha \approx 0.07$ in this experiment. The next two rows, Figure 2.6b and c, are experiments for the similar tilt angle $\alpha = 15°$ but varying inverse dimensionless height λ with a larger value in Figure 2.6b than in Figure 2.6c. Again, there are two vortices that appear to have centers that are close to the tank wall on either side. The

50 | 2 Fluid Stirring in a Tilted Rotating Tank

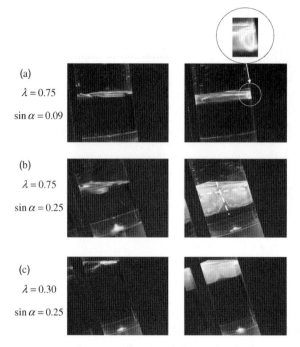

Figure 2.6 Experimental results for low Reynolds number stirring in a tilted-rotating tank for values of the tilt angle and dimensionless inverse liquid height, λ as indicated in the left-hand column. The inset at the top of the page shows a small vortex that forms near the tank wall. Final state images taken after approximately 10 000 s.

size (cross-sectional area) of the two mixed domains appear to be very close in value in these two experiments. There is a noticeable difference though in the details of the patterns that have formed in the cross sections. In Figure 2.6c the region near the liquid rotation axis appears clear and very visible. In Figure 2.6b though there appears to be some interaction between the liquid rotation axis and the bottom wall in the bulk flow due to the smaller fluid volume.

The next set of images, shown in Figure 2.7, show the transition from small to larger tilt angles for various values of λ as indicated. In each of these experiments the product of $\lambda \sin \alpha$ is in the range $0.1 < \lambda \sin \alpha < 0.4$ suggesting that the asymptotic parameter ε^2 is not too small and is finite. There is a noticeable difference in the shape of liquid rotation axis, which is clearly visible in each of the images. It increases in horizontal length as the tilt angle increases [48]. For example, the liquid rotation axis in Figure 2.7d and f look similar (location is indicated) where the values of λ are similar, despite the fact that the tilt angles are not. Also, when comparing experiments with different λ but similar tilt angle, there is little difference in the appearance of the mixed cross-sectional area. This is clearly seen when comparing Figure 2.7e and f.

In each of the images in Figure 2.7, the last set for the low Reynolds number experiments, the wall vortex occupies most of the flow domain suggesting that it is

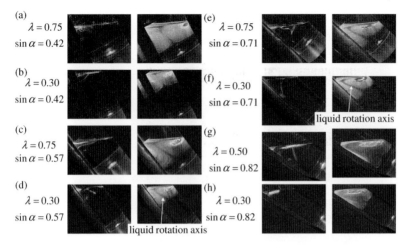

Figure 2.7 Series of experimental results for the tilted-rotating tank system with values of the tilt angle α and λ as indicated. Each series shows the initial and final condition. Final state images taken after approximately 10 000 s.

responsible for fluid stirring. However, additional information on fluid stirring is revealed by taking a closer look at Figure 2.7e and h in particular in the vicinity of the liquid rotation axis. In these two experiments there appears to be some evidence of periodic shearing in the form of small regions where the light intensity is brighter in, what appears to be, a periodic manner. This is the first evidence of periodic shearing in these images. Evidence is more robust when looking at time-lapsed sequences of images for a given experiment.

The last set of images for the low Reynolds number experiments, shown in Figure 2.8, continues to display the behavior described in the previous paragraph. Here the values of $\lambda \sin \alpha$ are large in either experiment shown, and the evidence of periodic shear driving the fluid stirring is clear in the vicinity of the liquid rotation axis. In fact, the influence of the wall vortex is very small and they do not appear near the wall (as indicated in the inset) and the two large stable regions, near the walls where no dye is present, exist for an extremely long period of time. The rate of mixing in this particular experiment also shows a decrease in the value when compared to the experiments performed at $\alpha = 55°$ [48].

2.3.2.2 Laminar Flow: Homogeneous Fluid

The next series of images, Figure 2.9, show results of stirring in a tilted rotating tank at laminar Reynolds numbers, $O(1)$. There are two values of λ and similar values for the tilt angle $\alpha = 25°$. The first column, Figure 2.9a–c, shows images of the results for the stirred cross-sectional area for $\lambda = 0.75$. At the lowest Reynolds number in the left column, Figure 2.9a, the torodial wall vortex is clearly visible. In the image below, at Reynolds number of 1.1, there is a secondary vortex that has formed near the bottom tank wall. The secondary vortex, in Figure 2.9b, is directly below the horizontal location of the torodial wall vortex in the image shown above, Figure 2.9a, where the

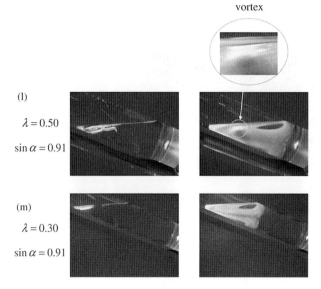

Figure 2.8 Final sequence of images for low Reynolds number stirring of fluid in a tilted-rotating tank. These final images are taken at the largest value of $\alpha = 65°$. The inset at the top shows the location of the vortex that appears to be detached from the wall and slowly drives a secondary flow that generates a large ordered region. Final state images taken after approximately 10 000 s.

experiment was performed at a lower value of the Reynolds number. As the Reynolds number increases further, Figure 2.9c, it appears that the vortex grows larger, or at least there is less stirring in the region below the torodial vortex, near the bottom tank wall, until the stirred region looks like a funnel in Figure 2.9d.

In the right column, Figure 2.9e–h, the tilt angle is the same as in the left column but the value of λ is small indicating a larger value for the liquid height. The image resulting from the experiment at the lowest value of the Reynolds number, Figure 2.9e, shows the torodial vortex near the wall and free surface. As the Reynolds number is increased by an order of magnitude, from Figure 2.9e and f, the stirred region appears to be stretched although it does not cover the complete cross-sectional domain. In fact the image for the $Re = 1.1$ experiment shows a region just below the stirred region where dye has been trapped (indicated in the image) and does not mix for the complete elapsed time of the experiment that is approximately 12 000 s. As the Reynolds number is increased from 1.1 to 2.0 the stirred region is larger and completely fills the cross-sectional area. However, as the Reynolds number is increased further, there is the appearance of interaction between the secondary vortices with the bottom wall. The mixed region at higher Reynolds numbers does not show the same conical vortex-type structure that is attached to the bottom wall as the same experimental at smaller height in this set of images. But there is evidence of this phenomenon occurring at these larger fluid volumes that has been previously investigated [53].

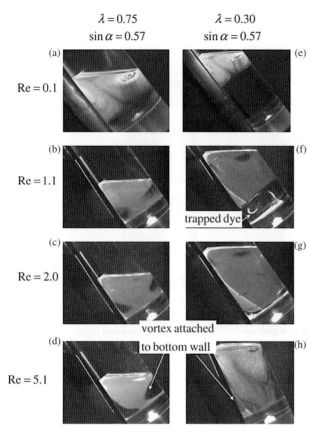

Figure 2.9 Qualitative comparison of fluid stirring in a tilted-rotating tank as the Reynolds number is increased for two different values of λ with approximately the same tilt angle $\alpha = 25°$. In the left-hand column, there is the appearance of ordered regions near the bottom tank wall that increases with increasing Reynolds number until the stirred region resembles a vortex that is attached to the bottom wall. A similar situation occurs on the right except there is some initial axial stretching of the stirred region at $Re = 1.1$ until the total domain is stirred at $Re = 2.0$ and then as the Reynolds number is increased to $Re = 5.1$ there is the appearance of the ordered region in a quasi-steady flow. (a) 10 000 s, (b) 10 000 s, (c) 1 000 s, (d) 2 000 s, (e) 10 000 s, (f) 12 780 s, (g) 7 800 s, and (h) 1 300 s.

2.3.2.3 Laminar Flow: Inhomogeneous Fluid

The next series of images are shown in Figure 2.10 and display some of the stirring results for inhomogeneous fluids. Each row of images show results for stirring using a given concentration of surfactant ϕ while the columns are arranged by the tilt angle and inverse-dimensionless height λ. The first column has images of results for $\lambda = 0.48$ with a tilt angle of $\alpha = 25°$. In this limit, there is little dispersion of the water as gravity and subsequent sedimentation of the dispersed phase fluid clearly dominates the stirring dynamics. In the next column where $\alpha = 55°$ there

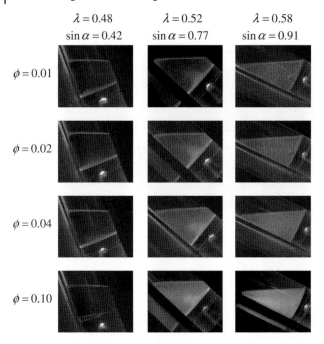

Figure 2.10 Examples of inhomogeneous fluid stirring in a tilted-rotating tank. For inhomogeneous fluid mixing, a surfactant is introduced into the system in order to help facilitate breakup of the drop. As the drops breakup into smaller and smaller diameter drops, their diameters become less than the laser width and the light is scattered, illuminating the drops. The left-hand column shows the concentration of surfactant (by volume) that is introduced into the dispersed phase fluid (water). The intensity of the light increases as the concentration of surfactant and the tilt angle are both increased. All images are taken after approximately 500 s.

appears to be more stirring of the two phases for a given concentration, and there is also an increase for increasing concentration with the other parameters fixed. There is a vortex that is visible in the second column results that resemble the vortices that appeared in the homogeneous laminar fluid flow experiments discussed in the previous paragraphs. The last column shows results for the largest tilt angle $\alpha = 65°$ where there is clearly more dispersion of the water phase for all concentrations.

At an angle of 55° there is little vertical dispersion of the droplets as they breakup. It seems to be independent of the drop size, since there appear to be many small droplets (as indicated by light intensity) even at the tilt angle of 25° seen in the first column for most of the concentrations shown. The liquid height must be reduced to a value that is less than the critical, that is, $\hat{V} < \hat{R}^3 \tan \alpha$ and then there is interaction between the bottom wall and free surface. This interaction is clearly responsible for the vertical displacement of the dispersed phase as evident for each concentration seen as the tilt angle increases from $\alpha = 55°$ to 65°.

2.3.3
Brief Discussion

A comparison between the theoretical analysis and experiments is presented in this section. In general, most of the main arguments of the asymptotic analysis appear to be verified by the experimental results. The appearance of a torodial vortex at the smallest tilt angle (Figure 2.6a) seem to reinforce the conclusion for the flow in the vicinity of the free surface that was drawn from the asymptotic analysis in the limit of the small tilt angle and large heights. It should also be noted that this small-wall vortex appears extremely close to the wall, and the passive dye takes a relatively long time to trace the flow in this experiment (Figure 2.6a). As the tilt angle is increased, see Figure 2.6b and c, the vortex appears to stir a larger volume of the fluid. The torodial vortex appears to be nearly symmetric about the liquid rotation axis, that appears as a region with no fluid stirring surrounded by the mixed fluid. In Figure 2.6c the values for the asymptotic parameters are $\varepsilon = 0.25$ based on $\sin\theta$ and $\varepsilon = 0.30$ based on λ. This was identified as the upper limit for the asymptotic analysis and appears to agree well with the experimental results shown in this figure where the stirred region occupies an equal vertical and horizontal region on either side of the liquid rotation axis. For values of $\sin\alpha > 0.3$ in the experimental data it appears that one vortex spans most of the stirred domains' cross-sectional area.

The laminar flow results seem to agree with the notion that larger Reynolds numbers do not necessarily lead to larger stirred regions of fluid. This is particularly true when looking at Figure 2.7 for the largest values of λ and as the Reynolds number is increased. Clearly, there is the appearance of a region near the bottom wall where the fluid does not appear to mix that grows with increasing Reynolds number. For the inhomogeneous fluid, the appearance of a vortex that transports the heavier drops to the bulk of the fluid is very similar to the vortices seen in the homogeneous laminar flow experimental results. The convective laminar flow is efficient for transporting nonneutrally buoyant species but only if the fluid volume is low enough so that the free surface intersects the bottom tank wall.

Overall, the tilted rotating tank geometry appears to efficiently stir homogeneous fluids near the free surface for low Reynolds numbers. At laminar Reynolds numbers, the flow exhibits stirring of a larger fluid domain although the convection generates a secondary torodial vortex that does not interact with the primary wall vortex and becomes a KAM surface.

2.4
Conclusion

The stirring and subsequent mixing of viscous fluids in a tilted-rotating tank has been studied both theoretically and experimentally. The theoretical analysis consisted of finding a reduced-order model by performing a power series expansion of the solid-body rotation in a tilted geometry that satisfies a minimal set of boundary condition for the flow of a liquid in a tilted rotating tank which are: (1) incompressibility,

(2) zero shear stress along the free surface, and (3) no penetration of the fluid through the free surface. Asymptotic analysis of the reduced-order model suggested that stirring for this flow is achieved though transport generated by a torodial vortex that emanates from the tank walls, and periodic shearing which occurs in the vicinity of a zero-flow region called the liquid rotation axis. The experiments consisted of using a passive dye to trace the flow fluid over the large number of rotation in an experimental Poincaré mapping of the flow field. The experiments verified many of the results of the asymptotic analysis.

Overall, the results suggested that fluid stirring in a tilted rotating tank may be useful in many processes where it is desirable to reduce the amount of energy that must incorporated into a system in order to achieve a certain quality of mixing. Unfortunately, the experiments suggested that the addition of convection, while reducing the amount of time needed to complete stirring (which then allows for diffusion to complete the mixing process), leads to the appearance of KAM surfaces which are barriers to mixing.

In the future, it will be beneficial to understand more complex behavior of the titled rotating tank mixer by injecting non-Newtonian fluid into the system or possibly using a polymer fluid for the continuous phase. Also, the effects of changing the tank diameter have not yet been investigated, which may lead to a richer variety of observed flow patterns.

Acknowledgments

The author would like to thank Asher Metchik who performed the low Reynolds number experiments, William Hourigan who performed the laminar flow Reynolds number experiments and Andrew White who performed the inhomogeneous laminar flow experiments. This work was supported in part by the NC Space Grant and a Duke Energy grant that supported Andrew White.

References

1 Alvarez-Hernández, M.M., Shinbrot, T., Zalc, J., and Muzzio, F.J. (2002) Practical chaotic mixing. *Chemical Engineering Science*, **57**, 3749–3753.

2 Aubin, J. and Xuereb, C. (2006) Design opf multiple impeller stirred tanks for the mixing of highly viscous fluids using CFD. *Chemical Engineering Science*, **61**, 2913–2920.

3 Campolo, M., Sbrizzai, F., and Soldati, A. (2003) Time-dependent flow structures and Langrangian mixing in Rushton-impeller baffled-tank reactor. *Chemical Engineering Science*, **58**, 1615–1629.

4 Chang, P.-K. and Peng, Y.-N. (2001) Influence of mixing techniques on properties of high performance concrete. *Cement and Concrete Research*, **31**, 87–95.

5 Jézéquel, P.-H. and Collin, V. (2007) Mixing of concrete or motars: dispersive aspects. *Cement and Concrete Research*, **37**, 1321–1333.

6 Kirca, O., Turanli, L., and Erdoğan, T.Y. (2002) Effects of retempering on consistency and compressive strength of concrete subjected to prolonged mixing. *Cement and Concrete Research*, **32**, 441–445.

7 Vickers, T.M., Farrington, S.A., Bury, J.R., and Brower, L.E. (2005) Influence of dispersant structure and mixing speed on concrete slump retention. *Cement and Concrete Research.*, **35**, 1882–1890.
8 Arnold, V.I. (1978) *Mathematical Methods of Classical Mechanics*, Springer, New York.
9 Angilelia, J.R. and Brancher, J.P. (2002) Note on chaotic advection in an oscillating drop. *Physics of Fluids*, **15** (1), 261–264.
10 Bajer, K. and Moffatt, H.K. (1990) On a class of steady confined Stokes flows with chaotic streamlines. *Journal of Fluid Mechanics*, **212**, 337–363.
11 Bajer, K. and Moffatt, H.K. (1992) Chaos associated with fluid inertia, in *Topological Aspects of the Dynamics of Fluids and Plasmas* (eds H.K. Moffatt, G.M. Zavlavsky, P. Comte, and M. Tabor), Kluwer, Dordrecht, The Netherlands, pp. 517–534.
12 Bajer, K., Moffatt, H.K., and Nex, F.H. (1990) Steady confined Stokes flows with chaotic streamlines, in *Topological Fluid Dynamics* (eds H.K. Moffatt and A. Tsinober), Cambridge University Press, Cambridge, pp. 459–466.
13 Bryden, M.D. and Brenner, H. (1999) Mass-transfer enhancement via chaotic laminar flow within a droplet. *Journal of Fluid Mechanics*, **379**, 319–331.
14 Cary, J.R., Escande, D.F., and Tennyson, J.L. (1986) Adiabatic-invariant change due to separatrix crossing. *Physical Review A*, **34** (5), 4256–4275.
15 Grigoriev, R.O. (2005) Chaotic mixing in thermocapillary-driven microdroplets. *Physics of Fluids*, **17** (3)
16 Kroujiline, D. and Stone, H.A. (1999) Chaotic streamlines in a steady bounded three-dimensional Stokes flow. *Physica D*, **130**, 105–132.
17 Mezić, I. (2001) Break up of invariant surfaces in action-action-angle maps and flows. *Physica D*, **154**, 51–67.
18 Neishtadt, A. (1984) The separation of motions in systems with rapidly rotating phase. *Journal of Applied Mathematics and Mechanics*, **48** (2), 133–139.
19 Ottino, J.M. (1990) Mixing, chaotic advection and turbulence. *Annual Review of Fluid Mechanics*, **22**, 207–253.
20 Rom-Kedar, V. and Poje, A.C. (1999) Universal properties of chaotic transport in the presence of diffusion. *Physics of Fluids*, **11** (8), 2044–2057.
21 Stone, H.A., Nadim, A., and Strogatz, S.H. (1991) Chaotic streamlines inside drops immersed in steady stokes flows. *Journal of Fluid Mechanics*, **232**, 629–646.
22 Vainshtein, D.L., Vasiliev, A.A., and Neishtadt, A.I. (1996) Changes in the adiabatic invariant and streamline chaos in confined incompressible Stokes flow. *Chaos*, **6** (1), 67–77.
23 Ward, T. and Homsy, G.M. (2001) Electrohydrodynamically driven chaotic mixing in a translating drop. *Physics of Fluids*, **13** (12), 3521–3525.
24 Ward, T. and Homsy, G.M. (2003) Electrohydrodynamically driven chaotic mixing in a translating drop. Part II: Experiments. *Physics of Fluids*, **15** (10), 2987–2994.
25 Lee, S.M., Im, D.J., and Kang, I.S. (2000) Circulating flows inside a drop under time-periodic non-uniform electric fields. *Physics of Fluids*, **12** (8), 1899–1910.
26 Neishtadt, A.I., Vainshtein, D.L., and Vasiliev, A.A. (1998) Chaotic advection in a cubic stokes flow. *Physica D*, **111**, 227–242.
27 Stone, Z.B. and Stone, H.A. (2005) Imaging and quantifying mixing in a model droplet micromixer. *Physics of Fluids*, **17** (1)
28 Aref, H. (1984) Stirring by chaotic advection. *Journal of Fluid Mechanics*, **143**, 1–21.
29 Kronig, R. and Brink, J.C. (1949) On the theory of extraction from falling droplets. *Applied Scientific Research*, **A2**, 142–154.
30 Strook, A.D., Dertinger, S.K.W., Ajdari, A., Mezić, I., Stone, H.A., and Whitesides, G.M. (2002) Chaotic mixer for microchannels. *Science*, **295**, 647–651.
31 Cortelezzi, L., Adrover, A., and Giona, M. (2008) Feasibility, efficiency and transportability of short-horizon optimal mixing protocols. *Journal Of Fluid Mechanics*, **597**, 199–231.
32 D'Alessandro, D., Dahleh, M., and Mezić, I. (1999) Control of mixing in fluid flow: A maximum entropy approach. *IEEE Transactions on Automatic Control*, **44** (10), 1852–1863.

33 Franjione, J.G. and Ottino, J.M. (1992) Symmetry concepts for the geometric analysis of mixing flows. *Philosophical Transactions of the Royal Society of London*, **338** (1650), 301–323.

34 Taylor, G.I. (1923) Stability of a viscous liquid contained between two rotating cylinders. *Philosophical Transactions of the Royal Society of London, Series A*, **223**, 289–343.

35 Wendl, M.C. (1999) General solution for the Couette flow profile. *Physical Review E*, **60** (5), 6192–6194.

36 Lichtenburg, A.J. and Lieberman, M.A. (1992) *Regular and Chaotic Dynamics*, Springer, New York.

37 Johnson, R.E. (1988) Steady-state coating flows inside a rotating horizontal cylinder. *Journal of Fluid Mechanics*, **190** (1), 321.

38 King, A.A., Cummings, L.J., Naire, S., and Jensen, O.E. (2007) Liquid film dynamics in horizontal and tilted tubes: Dry spots and sliding drops. *Physics of Fluids*, **19**, 042102.

39 Landau, L. and Levich, B. (1942) Dragging of a liquid by a moving plate. *Acta Physiochim. (USSR)*, **17**, 42–54.

40 Ruschak, K.J. (1985) Coating Flows. *Annual Review of Fluid Mechanics*, **17**, 65–89.

41 Ruschak, K.J. and Scriven, L.E. (1976) Rimming flow of liquid in a rotating horizontal cylinder. *Jounrla of Fluid Mechanics*, **76**, 113–125.

42 Thoroddsen, S.T. and Mahadevan, L. (1997) Experimental study of coating flows in a partially-filled horizontally rotating cylinder. *Experiments in Fluids*, **23**, 1–13.

43 Taylor, G.I. (1934) The formation of emulsions in definable fields of flow. *Proceedings of Royal Society of London A*, **138**, 501–523.

44 Tjahjadi, M. and Ottino, J.M. (1991) Stretching and breakup of droplets in chaotic flows. *Journal of Fluid Mechanics*, **232**, 191–219.

45 Tjahjadi, M., Ottino, J.M., and Stone, H.A. (1992) Satellite and subsatellite formation in capillary breakup. *Journal of Fluid Mechanics*, **243**, 297–317.

46 Fountain, G.O., Khakhar, D.V., and Ottino, J.M. (1998) Visualization of three-dimensional chaos. *Science*, **4281**, 683–686.

47 Fountain, G.O., Khakhar, D.V., Mezić, I., and Ottino and, J.M. (2000). Chaotic mixing in a bounded three-dimensional flow. *Journal of Fluid Mechanics*, **417**, 265–301.

48 Ward, T. and Metchik, A. (2007) Viscous fluid mixing in a tilted tank by periodic shear. *Chemical Engineering Science*, **62** (22), 6274–6284.

49 Moffatt, H.K. (1964) Viscous and resistive eddies near a sharp corner. *Journal of Fluid Mechanics*, **18**, 1–18.

50 Sanz-Serna, J.M. and Calvo, M.P. (1994) *Numerical Hamiltonian Problems*, Chapman and Hall, London.

51 Shankar, P.N. and Kumar, M. (1994) Experimental determination of the kinematic viscosity of glycerol water mixtures. *Proceedings of Royal Society of London A*, **444**, 573–581.

52 Ernst, R.C., Watkins, C.H., and Ruwe, H.H. (1936). The physical properties of the ternary system ethyl alcohol glycerin water. *Journal of Physical Chemistry*, **40** (5), 627–635.

53 Ward, T. and Hourigan, W. (2009) Experimental investigation of transition to laminar mixing of a homogeneous viscous liquid in a tilted-rotating tank. *Chemical Engineering Science*, **64** (23), 4919–4928.

3
Lagrangian Coherent Structures
Shawn C. Shadden

3.1
Introduction

Mounting evidence suggests that fluid advection can effectively be studied by considering special material surfaces, which are referred to here as Lagrangian coherent structures (LCSs).[1] What makes these material surfaces special is their distinguished attracting or repelling nature. Notably, LCS are often locally the most strongly attracting or repelling material surfaces in the flow, and as such have a strong influence on the flow topology. In fact, by understanding their evolution, one can often reveal *mechanisms* that underly complex laminar, and even turbulent, fluid transport in conspicuous detail.

Conceptually, LCS can be approached from the dynamical systems perspective, or from a more physically based fluid mechanics perspective. Starting from the former, we note that fluid advection is described by the equation

$$\dot{\mathbf{x}}(\mathbf{x}_0, t_0, t) = \mathbf{u}(\mathbf{x}, t) \qquad (3.1)$$

where $\mathbf{u}(\mathbf{x}, t)$ is the velocity field of a fluid and $\mathbf{x}(\mathbf{x}_0, t_0, t)$ describes the motion (trajectory) of a fluid element, or equivalently a *material point*,[2] starting at position $\mathbf{x}_0$ at time t_0; we assume nominally volume-preserving flow, $\nabla \cdot \mathbf{u} \approx 0$. Since the motion of fluid is, generally speaking, chaotic, revealing salient flow features helps us understand how the flow is organized. Indeed, the study of *coherent* or *organizing* structures in fluid mechanics has surely been of interest for as long as we have

---

1) Herein we use LCS to abbreviate both singular and plural forms.

2) The concept of a fluid element is an idealization – but one that is overwhelmingly used in modeling fluid mechanics, for example, the Navier-Stokes equation and resulting solution are based on this assumption. Resulting flow data is typically not "amended" by diffusion for LCS computations, as LCS targets advection, which indeed is the primary mode of transport over the length and time scales typically considered. For many applications though, interest lies in the advection of matter that is approximately transported by the fluid (e.g., bubbles, aerosols, suspensions, and emulsions); in such cases, the dynamics (Eq. (3.1)) can be appropriately augmented [5], or LCS of the fluid can be seen as approximate LCS for the advected matter, and under appropriate conditions, their relevancy can be established rigorously [6, 7], see also [8,9].

Transport and Mixing in Laminar Flows: From Microfluidics to Oceanic Currents,
First Edition. Edited by Roman Grigoriev.
© 2012 Wiley-VCH Verlag GmbH & Co. KGaA. Published 2012 by Wiley-VCH Verlag GmbH & Co. KGaA.

studied fluid motion. In more recent history, the field of dynamical systems has assisted in making this effort precise. Generically, Eq. (3.1) is a nonlinear ordinary differential equation, which is the focus of study in dynamical systems theory. Hyperbolic fixed points and their associated stable and unstable manifolds[3,4] are known to be organizing structures in dynamical systems, at least in the more mathematically tractable problems traditionally considered in that field [1], as described below. Briefly, in steady 2D flow, stable and unstable manifolds of hyperbolic fixed points behave as separatrices, partitioning regions with similar dynamics and providing the skeletal structure of the flow topology. Steady 2D flow is exceptional though, because it is integrable. Conversely, time-dependent and/or 3D flow is typically not integrable, and fluid motion generically is chaotic, making understanding the flow topology exponentially harder. The simplest unsteadiness that can be added to a system is a periodic perturbation, yet this simple change in the velocity field often translates to complex change in the fluid motion, that is, chaos. In such systems, periodic trajectories can be seen as fixed points using a stroboscopic view (Poincare map) and the stable and unstable manifolds of these fixed points can reveal a template organizing the fluid's chaotic motion. Therefore, computing these manifolds is central to understanding transport in periodic systems, and indeed this aspect of chaotic advection has become well-developed, see, for example, [2–4].

In practical applications, fluid flows span a spectrum of unsteadiness. As complexity increases, the motivation for identifying organizing structures becomes increasingly compelling, yet applying invariant manifold theory in complex flow problems is challenging, for example, it is difficult to even *define* these concepts satisfactorily for temporally aperiodic flows, which is well-documented [10–12]. Hence, the definitions for, and relevancy of, stable and unstable manifold theory in systems with general time dependence are not obvious, and research in this area continues to mature. At least in the sense presented here, these manifolds have fallen under the broader category LCS – a terminology adopted from Haller and Yuan [13]. This terminology helps broaden our description, and categorization, of organizing structures from the confines of the traditional definitions of stable and unstable manifolds, while maintaining a practical methodology for revealing "effective" stable and unstable manifolds in flows with arbitrary time dependence.

From the fluid mechanics perspective, fluid has a fascinatingly complex behavior and understanding the nature of this complexity is one of the great challenges in fluid mechanics. Growth in technology has enabled tremendous computational and empirical capabilities to derive velocity field data in wide-ranging applications. However, obtaining data is certainly not enough, we must be able to properly

---

3) A manifold can be viewed herein as a smooth surface, and it is tacitly assumed that a surface is meant herein to mean hypersurface (i.e., for 2D systems, surface is synonymous with curve).

4) For 2D systems, eigenvalues can be purely imaginary and, thus, the velocity gradient "nonhyperbolic." In 3D, since complex eigenvalues occur in conjugate pairs, this requires one, or all three, eigenvalue to remain zero. While this is certainly possible, it is presumably less common over finite time intervals in complex flows (in the interior at least, since no-slip boundaries are clearly nonhyperbolic).

assimilate the complex information encoded in that data. Comprehending the motion of the fluid is paramount in studying fluid mechanics but often precise understandings are elusive due to the inherent complexity of fluid flow. As described in Ref. [14], "human temporal perception struggles with untangling chaotic trajectories in a turbulent flow." Thus, we are tempted to more tractable descriptions, based on instantaneous rate-of-change information, such as the velocity field or related Eulerian measures. However, the relevancy of these fields, or coherent structures derived from them, to the actual fluid motion is dubious since time dependence can "wreak havoc" on conclusions drawn from instantaneous information. Furthermore, time series of instantaneous fields frequently mislead our perception from the reality of the integrated behavior of the fluid motion over time. These troubles are well-documented, but perhaps not widely appreciated. On the other hand, Lagrangian-based measures that explicitly track the fluid motion are less common, and in many cases *ad hoc* or qualitative. The advantage of the methods presented herein is the ability to systematically and succinctly encode key Lagrangian information into a single field from which we can define structures that remain relevant in space and time, are coordinate-frame invariant, and perhaps most importantly convey information regarding fundamental *mechanisms* of fluid transport.

The structures we compute are termed *Lagrangian* because they are defined from the fluid motion, as opposed to an instantaneous Eulerian snapshot, and because they are themselves material surfaces advected by the flow. They are also termed *coherent* because they have distinguished stability compared to nearby material surfaces, and consequentially, LCS can often be identified with familiar coherent flow features. For example, the coherent patterns traced out by visual markers (e.g., dye) we see emerge when visualizing fluid flow are often manifestations of underlying LCS. The Lagrangian nature of LCS makes them "transport barriers;" however, it is their distinguished stability, which often translates into separatrix behavior, that substantiates their importance. In this chapter, we review and summarize the main concepts underlying this approach to studying fluid advection, and how these structures are commonly computed in practice and related practical concerns. For underlying mathematical details, or specific application-oriented computational results, we have attempted to provide appropriate references.

3.2
Background

Let us revisit the nature of hyperbolic fixed points in steady flow for developing intuition regarding LCS and to make ideas introduced in the section above more concrete. Consider the steady vortex dipole shown in Figure 3.1a – a pedagogical application that we will build upon. Vortices are considered the building blocks for turbulence, and thus understanding their structure is fundamental to fluid mechanics. In the vortex dipole, there are two hyperbolic saddle-type fixed points. Their stable/unstable manifolds are the trajectories that asymptote to these points in forward/backward time. In general, stable and unstable manifolds act as separatrices,

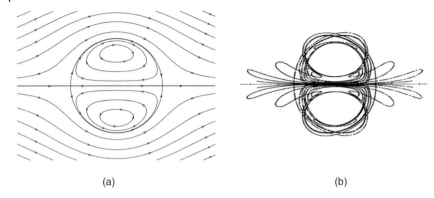

(a)   (b)

Figure 3.1 Panel (a): Streamlines for a steady vortex dipole. Two hyperbolic saddle-type stagnation points are located upstream and downstream. Their stable and unstable manifolds form so-called *heteroclinic* connections between the saddle points, and behave as *separatrices* partitioning the flow into regions of distinct dynamics—a common property of stable and unstable manifolds. Panel (b): Periodically perturbed vortex dipole. Heteroclinic tangle formed by transverse intersection of stable and unstable manifolds when a perturbation is applied to the steady vortex dipole, reproduced from [15] with permission.

partitioning regions of different dynamics. This behavior is exemplified here as these manifolds define the vortex boundaries, separating recirculating fluid from non-recirculating fluid. In general, hyperbolic fixed points, which are required to be saddle-type due to incompressibility of fluid, form permanent hubs, drawing in fluid and expelling the fluid in diverging directions. Namely, the saddle points in the vortex diapole pull in the fluid along the stable manifolds and expel the fluid in diverging directions along the unstable manifolds. Restating, the fixed points are hubs directing the flow, and the invariant manifolds delineate how fluid will be directed by those hubs. This behavior is central to the importance of stable/unstable manifolds, and as it turns out, this behavior remains relevant in complex unsteady flows; however, the role of fixed points is often replaced by appropriately behaved moving trajectories.

The clean partitioning of dynamically distinct regions in the domain by stable and unstable manifolds, and lack of mixing, as seen in Figure 3.1a, occurs primarily in 2D time-independent problems. The geometrical structure of the transport topology is usually much more complicated when systems become time-dependent or 3D. When steady 2D systems, as in Figure 3.1a, are subjected to periodic perturbations, the fixed points typically perturb to periodic orbits, and their associated stable and unstable manifolds transform from neat heteroclinic connections to complex heteroclinic tangles, as illustrated in Figure 3.1b, leading to the theory of lobe dynamics, see, for example, [15, 16]. This transverse intersecting of manifolds is one of the hallmarks of chaos. If one visualizes the chaotic dynamics along with these manifolds, the dynamics are revealed to be, in fact, rather orderly and predictable. A great deal of developments have been made in understanding chaotic dynamics in periodically, and quasi-periodically, perturbed systems using invariant manifold theory, see, for example, [2–4]. Notably, such information has proved very valuable in the understanding, and practical design, of efficient fluid mixing.

Several difficulties exist for computing stable and unstable manifolds in nonperiodic systems. First, it is not immediately clear what one should compute the invariant manifolds of. For steady flow, hyperbolic fixed points and for periodic flow, hyperbolic periodic trajectories are the *persistent hubs* organizing the flow topology. However, fixed or periodic trajectories are not commonly encountered in most practical applications due to the general unsteadiness of most flows. Furthermore, trajectories that are hyperbolic at one instant might quickly change their stability – something that is not a problem when working with fixed points in steady and periodic systems. Next, the very definition of stable/unstable manifolds becomes unwieldily in the "real world." Defining stable and unstable manifolds requires asymptotic limits [1, 17]. However, such limits are irrelevant for practical applications since (1) the fluid's motion is known only from an inherently finite data set, for example, from a computational fluid dynamics simulation or empirical measurement, and (2) since trajectories can gain or loose hyperbolicity over time and we desire to understand inherently transient phenomena. Amending invariant manifold theory as needed to rigorously and unambiguously establish and compute stable and unstable manifolds of hyperbolic trajectories in nonperiodic, finite-time systems takes care (see, e.g., [11, 18, 19]). Herein, we take a "less rigorous" approach that seeks to uncover such manifolds, and perhaps more general organizing structures, using hyperbolicity measures, such as the finite-time Lyapunov exponent (FTLE) distribution, as described in Section 3.3. Numerous and wide-ranging applications have demonstrated, indeed, that structures analogous to stable and unstable manifolds are effectively determined in nonperiodic, and even turbulent, flow, and in fact this approach has revealed that such structures are seemingly ubiquitous in fluid mechanics. LCS have garnered much attention because they often provide a more precise (and temporally relevant) depiction of common fluid mechanics constructs (e.g., vortices, flow separation, and stirring) than prevailing methods of flow characterization, making them very useful indeed. In the context of mixing, while lobe dynamics generally becomes less common in nonperiodic systems, LCS continue to control the stretching, folding, and alignment mechanisms underlying kinematic mixing.

We note that the term LCS is used in various communities to mean different things. An important distinction is that LCS is used here to describe coherent *surfaces*, and not coherent *regions* in the flow. Nonetheless, often due to their separatrix nature, such LCS end up identifying boundaries to regions of coherent dynamics, as stable/unstable manifolds often do (cf. Figure 3.1, see also Refs. [20, 21] for comparison with boundaries to almost-invariant regions). In theory, LCS encompasses the notion of stable and unstable manifolds of hyperbolic trajectories, and in practice, as often computed, typically reveal these manifolds in systems where they are known to exist. However, it is not clear that we require that LCS be stable/unstable manifolds, at least in a traditional sense. Indeed, it is perhaps advantageous to think of an LCS, defined by Haller [22], more broadly as a locally strongest repelling or attracting material surface.

3.3
Global Approach

Considering LCS as the most repelling or attracting material surfaces we could ask what is the best way to compute them. Testing the expansion or contraction about every material surface in the flow is hopeless, since there is an infinite possibility of surfaces to test and there is no good way to discretize the set of all possible surfaces. Instead, we take a more indirect approach. We discretize the fluid domain with a dense grid of material points, measure the Lagrangian expansion rate (roughly "hyperbolicity") about each material point, plot the spatial distribution, and extract surfaces that maximize the measure. This is essentially how *repelling* LCS are most commonly computed. *Attracting* LCS are analogously computed by reversing time, as expansion in backward time implies contraction in forward time. Since the computation is so similar, we often do not distinguish the direction of integration in discussions of methodology. One may contrast this "global approach" to how stable/unstable manifolds are traditionally computed, whereby the computation begins with the identification of a *specific* hyperbolic trajectory and stable and unstable manifolds are grown from the stable and unstable subspaces of that trajectory. The global approach proceeds independent of any specific hyperbolic trajectories, whereby the distribution of an appropriate hyperbolicity measure generally reveals *all* influential finite-time hyperbolic structures (e.g., the relevant hyperbolic trajectories and their associated stable and unstable manifolds).

One should keep in mind that when we speak of attracting and repelling surfaces, we do so in the context of nominally volume-preserving fluid flows. Thus, material points about a repelling LCS tend to compress in the tangential direction and expand in the normal direction. Likewise, material points about an attracting LCS will tend to expand along the LCS and contract in the normal direction toward the structure.

The global approach was built on the observation that material points straddling stable/unstable manifolds typically separate faster in forward/backward time than pairs of points not straddling such manifolds. Alternatively, as previously stated, LCS can more generally be defined as material surfaces with a most repelling or attractting property, without regard to them being stable or unstable manifolds of something.[5] This strategy for computing LCS owes to many people. Bowman wrote a paper [23] that proposed the use of a finite-strain (FS) field for locating stable and unstable manifolds. As summarized by Jones and Winkler [24], Bowman's "idea is to infer the manifold geometry by considering the stretching associated with the hyperbolicity of these structures." Bowman's paper captured the essence of how LCS are often computed nowadays (this paper was rejected from publication yet has been numerously acknowledged). As noted by Bowman, the computation of the FS field is similar

---

5) Stable/unstable manifolds do not always produce ridges in the FTLE field, see, for example, [34, 40, 41]. This has mainly been demonstrated in simple flows, as noted in [41], that lack sufficient spatial heterogeneity or with "uniform" chaos. That is, while stable and unstable manifolds usually act as separatrices, they are not necessarily the most repelling or attracting surfaces. Recently, Haller [22] introduced the notion of "weak" LCS and how to "constrain" the computation to distinguish the invariant manifolds.

to the computation of the FTLE field, which had been used in 1991 by Pierrehumbert [25] and in 1993 by Pierrehumbert and Yang [26] to reveal "structure" in atmospheric flows. Doerner *et al.* [27] published a paper that argued that local maxima of the FTLE field coincide with stable manifolds of hyperbolic fixed points (focussing on steady systems). Winkler [28] noted that such measures are closely related to measures of relative dispersion (RD), especially in practice, which, for example, von Hardenberg *et al.* [29] also utilized, along with FTLE fields, as a means to locate invariant manifold-type structures in geophysical flows. More precise quantification of the FTLE approach was described by Haller [30, 31], followed by Shadden *et al.* [32] and Lekien *et al.* [33]. Haller and Yuan [13] also proposed the hyperbolicity time (HT) measure that was more specific in its measure of hyperbolicity than existing "finite-stretching" measures, and Haller's papers [13, 30, 34, 35] developed a more rigorous foundation to LCS compuation than previously available; indeed, terminology and basic ideas we discuss here were significantly influenced by these fundamental works. The similar, but distinct, finite-size Lyapunov exponent (FSLE), developed by Aurell *et al.* [36], which measures time to separate a set distance instead of the FTLEs measure of distance separated in a set time, was also proposed [37–39] as a suitable measure in practice.

3.3.1
FTLE

So far, the spatial distribution of the FTLE appears to be an effective way to detect LCS computationally; in fact, it is often considered the "de facto" method. For illustration purposes, the application of this approach to a more practical version of the vortical flow considered above is shown in Figure 3.2. Briefly, planar velocity data was measured surrounding a mechanically generated vortex ring using particle image velocimetry (PIV). The velocity data was used to compute trajectory data, and subsequently FTLE fields in both forward and backward time (FTLE computation is described in more detail in Section 3.4). Snapshots of the forward and backward FTLE fields, and the instantaneous PIV velocity field, are plotted at one time instant in Figure 3.2. Visually, the FTLE fields reveal distinct curves of high FTLE; these curves

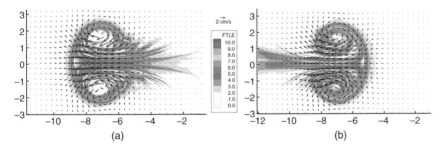

Figure 3.2 Backward and forward time FTLE fields, panels (a) and (b) respectively, for empirical vortex ring reveal attracting/repelling LCS that define vortex boundary and control transport and mixing. For further details see [42, 43].

are LCS analogous to the stable and unstable manifolds shown in Figure 3.1b (unlike the computation of the manifolds in Figure 3.1b, the LCS computation proceeds regardless of the time dependence of $u(x,t)$, making it applicable in more general settings). Together, these LCS reveal the *precise* vortex boundary, and precisely how fluid is entrained/detrained to/from the vortex ring as it forms and propagates; highly relevant information that is notably absent when employing traditional analyses techniques (e.g., velocity field, vorticity, streamlines, and Q-criterion).

Lyapunov exponents, in general, have long been used to determine the predicability, or sensitivity to initial conditions in dynamical systems. The FTLE measures the separation rate between initially close trajectories. Roughly, if two material points are initially separated by a small distance $|\xi_0|$ at time t_0, then the separation at some later time will be

$$|\xi_t| \approx e^{\Lambda(t-t_0)}|\xi_0| \qquad (3.2)$$

or equivalently

$$\frac{|\xi_t|}{|\xi_0|} \approx e^{\Lambda(t-t_0)} \qquad (3.3)$$

where Λ denotes the FTLE, which is a function of space, time and integration length. To see this we first note that the precise separation between two arbitrary material points starting at time t_0 from positions x_0 and $x_0 + \xi_0$ is given by

$$\xi_t = F_{t_0}^t(x_0 + \xi_0) - F_{t_0}^t(x_0) \qquad (3.4)$$

where $F_{t_0}^t : x_0 \mapsto x(x_0, t_0, t)$ denotes the flow map. By expanding the function $F_{t_0}^t(x_0 + \xi_0)$ in ξ_0 in Eq. (3.4), the distance between the two points over time is

$$|\xi_t| = \left|\nabla F_{t_0}^t(x_0) \cdot x_0\right| + \mathcal{O}(|\xi_0|^2) \qquad (3.5)$$

which, by definition of the Euclidean norm, can be rewritten as

$$|\xi_t| = \sqrt{e^\top \cdot \nabla F_{t_0}^t(x_0)^\top \cdot \nabla F_{t_0}^t(x_0) \cdot e} \, |\xi_0| + \mathcal{O}(|\xi_0|^2) \qquad (3.6)$$

where $e = \xi_0/|\xi_0|$ and $\top$ denotes transpose. The stretching factor, or coefficient of expansion, between the two points becomes

$$\lim_{|\xi_0| \to 0} \frac{|\xi_t|}{|\xi_0|} = \sqrt{e^\top \cdot \nabla F_{t_0}^t(x_0)^\top \cdot \nabla F_{t_0}^t(x_0) \cdot e} \qquad (3.7)$$

when $|\xi_0|$ becomes small. Letting $\lambda^i(x_0, t_0, t)$ denote the ith eigenvalue, numbered in decreasing order, of the right Cauchy–Green strain tensor

$$C(x_0, t_0, t) = \nabla F_{t_0}^t(x_0)^\top \cdot \nabla F_{t_0}^t(x_0) \qquad (3.8)$$

and $e^i(x_0, t_0, t)$ the associate eigenvectors, the coefficients of expansion in the eigenvector directions, that is, $\xi_0 = |\xi_0|e^i$, are given by

$$\lim_{|\xi_0|\to 0}\frac{|\xi_t|}{|\xi_0|}=\sqrt{\lambda^i(\xi_0,t_0,t)} \tag{3.9}$$

Note, $\sqrt{\lambda^i}=\sigma^i$ is the singular value of the deformation gradient $\nabla F_{t_0}^t(x_0)$. The FTLEs are defined by taking the natural logarithm of the coefficients of expansion, and dividing by the length of time the expansion occurred over (*integration time*) to get the average logarithmic expansion rate

$$\Lambda^i(x_0,t_0,t)=\frac{1}{|t-t_0|}\ln\sqrt{\lambda^i(x_0,t_0,t)}=\frac{1}{t-t_0}\ln\sigma^i(x_0,t_0,t) \tag{3.10}$$

The largest FTLE, Λ^1 is often referred to without distinction as the FTLE, Λ.

The eigenvectors e^i form an orthonormal basis. Nearly all generic perturbations ξ_0 will have a component in the e^1 direction, and growth of the perturbation will eventually be dominated by growth of that component since it has the largest eigenvalue. This observation, along with the proceeding derivations, justifies Eq. (3.3).

We last note that geometrical features of the fields $\lambda(x_0,t_0,t)$, $\sigma(x_0,t_0,t)$, or $\Lambda(x_0,t_0,t)$ are roughly equivalent since the mappings between these measures (Eq. (3.10)) are monotonic. For example, local extrema of one field will correspond to local extrema of the other. However, because of the often exponential divergence of particles straddling LCS, the logarithmic-based FTLE field can produce crisper structures. For most purposes, these fields, and *ad hoc* measures introduced above, can be used interchangeably, or with minor modification, in the discussions that follow. We focus on FTLE because it has been most widely used, and more rigorous results exist for this method [22, 31–33, 35].

3.3.2
FTLE Ridges

LCS appear as local maximizing surfaces of FTLE. In some ways, it is striking that when one plots a suitable scalar measure of hyperbolicity, such as the FTLE, one can clearly reveal an otherwise hidden template of transport. That is, why should we expect that if we plot the spatial distribution of the FTLE, it will form well-defined surfaces and not be more diffuse? Why should we expect these surfaces to be influential separatrices, or for them to be transported by the flow? Perhaps equipped with the right perspective such behavior seems sensible and justifiable, but it is not obvious such coherence will persist in the complexity of turbulent flow, especially when employing *averaged* stretching statistics such as the FTLE. Clearly, material surfaces with distinguished stability persist in all sorts of applications (cf. Section 3.6) and are readily detectable using suitable measures such as the FTLE.

Shadden *et al.* [32] formalized the definition of LCS as ridges of the FTLE field, an idea introduced in [31], and that definition was extended to higher dimensional systems in Lekien *et al.* [33]. These papers helped to make this approach precise, and consequently to help better derive properties of "ridges" of FTLE and to form a basis for computational strategies. In [32], we defined a ridge as essentially a curve on which the FTLE is locally maximized in the transverse direction, leading to a "second-derivative ridge" definition that required that a ridge was a curve (more generally, hypersurface in high dimensions) where the first derivative of the FTLE field must be zero in the normal direction, n, and the second derivative of the field must be negative and minimum in the normal direction. Furthermore, we desired hyperbolic hypersurfaces where only the largest Lyapunov exponent be positive, with the rough idea being that there should be attraction within the surface and repulsion normal to it. These conditions for 3D systems can be stated mathematically as

1) $\nabla \Lambda^1 \cdot n = 0$
2) $\langle n, \nabla^2 \Lambda^1 n \rangle = \mu_{\min}(\nabla^2 \Lambda^1) < 0$
3) $\Lambda^1 > 0 > \Lambda^2, \Lambda^3$

where $\mu_{\min}(\nabla^2 \Lambda^1)$ is the minimum eigenvalue of $\nabla^2 \Lambda^1$.

As the computation of FTLE/LCS matured, deficiencies were noted, see, for example, [32, 34, 40, 41] and discussion in section below. To address these issues, Haller [22] more recently derived necessary and sufficient conditions for LCS in terms of invariants of the strain tensor. Recalling that the FTLE is derived from the largest eigenvalue of the finite-time strain tensor, this approach can be seen as taking a step back to the more complete stretching information encoded in $C(x_0, t_0, t)$. A primary motivation was to ensure that LCS are locally the *most normally* repelling structures (as motivated below). Those conditions can subsequently be used to refine the FTLE ridge definition; namely, sufficient conditions that can be checked are that $e_1 = n$ and $|\nabla e_1 \cdot e_1| \leq 1$, and

$$\langle n, \nabla^2 \lambda^1 n \rangle < -2\lambda_1 |\nabla e_1 \cdot e_1| \left(\frac{\lambda_1}{\lambda_2} + \frac{\lambda_1}{\lambda_3} \right)$$

note this last condition is in terms of λ_is, which can be rewritten in terms of Λ_is via Eq. (3.10). Under these conditions, condition 3 above can be relaxed to $\Lambda^2 \neq \Lambda^1 > 0$. An important condition for normal hyperbolicity is $e_1 = n$; interestingly, it turns out that $e_1 \to n$ for persistent LCS, that is, the direction of dominant expansion/contraction converges to the normal direction.

The above mathematical criteria can be taken as a working definition of LCS, but we note that LCS are physical objects in the flow, and as such exist independent of any metric used to compute them. Nonetheless, mathematically precise definitions, as given above, are often necessary, if higher level definitions are not amenable to rigor or algorithmic implementation.

3.3.3
Nature of Stretching

The study of stretching in fluid flow has a long history, which we cannot possibly overview here. However, we do want to point out some pertinent comments

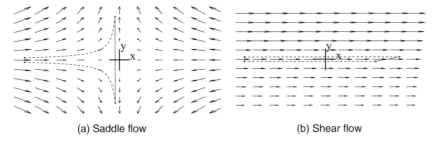

Figure 3.3 Stretching about a material line can be due to normal expansion (a) or tangential expansion (b). Certain stretching diagnostics used for LCS detection (e.g., FTLE) can obscure this distinction.

regarding stretching and hyperbolicity. A common physical interpretation of FTLEs regards the deformation of a spherical fluid element. Based on the arguments above, an *infinitesimal* sphere of fluid will deform into an ellipsoid when advected by the flow. The principal axes of the ellipsoid align with the eigenvectors of the strain tensor (Eq. (3.8)), and each principal axis is equal to the original radius of the sphere scaled by the corresponding coefficient of expansion. Thus, the FTLEs represent the average logarithmic expansion rates of each principal axis. Note for volume-preserving flow, the sum of Lyapunov exponents must be zero.

Hyperbolicity is often synonymous with its manifestation, stretching, in fluid mechanics. In a broad sense, a trajectory is hyperbolic if infinitesimal perturbations to that trajectory expand or contract over time. How this should be interpreted in finite time is not obvious. Trajectories can change behavior over time, and long-term limits cannot be computed, or are perhaps irrelevant since we may be interested in transient behavior. The physical interpretation in the proceeding paragraph motivates defining finite-time hyperbolicity in terms of FTLEs.[6] For example, Wiggins [44] defines a trajectory as hyperbolic over a finite time interval if none of its FTLEs is zero. Correspondingly, a case could be made that the semimajor axis, or equivalently the largest FTLE, can be viewed as a *measure* of hyperbolicity for that time interval since nearly all perturbations approach this expansion rate (Eq. (3.3)).

This interpretation of hyperbolicity is a bit generic though, because almost all trajectories are hyperbolic under such a definition. Also, while the FTLE, or more generically any time-averaged Lagrangian measure of stretching, will indicate levels of "hyperbolicity" for a trajectory, such measures can obscure the *nature* of the stretching. A revealing juxtaposition is shown in Figure 3.3. In both panels (a) and (b), the *x*-axis is a material surface. In panel (a), we see growth of a line element straddling the axis in the normal direction for a flow field topologically equivalent to flow near a hyperbolic trajectory. Alternatively, in panel (b), we see

---

6) Hyperbolicity has also been phrased in terms of "exponential dichotomy" [45].

growth of two points straddling the *x*-axis in the tangential direction for a shear flow.[7] In Figure 3.3a, the *x*-axis is a stable manifold and the general stretching behavior observed is heuristically what one typically associates with an LCS. However, locations of high shear certainly can cause high values of FTLE, and surfaces of maximal shearing rate may lead to ridges in the FTLE (or similar) field. This paradox was noted in [22, 31], with the shear profile shown in Figure 3.3b taken from [22]

$$\dot{x} = 2 + \tanh(y)$$
$$\dot{y} = 0 \tag{3.11}$$

For this flow, the shear rate is locally maximized at $y = 0$ and thus the FTLE field would admit a ridge along the *x*-axis. However, it would seemingly be awkward to refer to the *x*-axis in Figure 3.3b as an LCS since it does not act as a stable manifold, or more generally a separatrix.

One can exclude such maximal shear surfaces if one is more choosy in defining a hyperbolic surface. The importance of normally hyperbolic invariant manifolds has been noted [17] and similar ideas have been employed for identifying LCS with *normally* hyperbolic material surfaces. Normal hyperbolicity can be defined in multiple ways in finite time. For example, as presented in Haller and Yuan [13], one could require that normal perturbations to the surface *continuously* expand/contract in the normal direction over the complete time interval of interest, leading to, for example, the HT approach to LCS detection [13, 30]. More recently, Haller [22] required that at the end of the time interval of interest, a perturbation that started normal to the surface must have expanded/contracted *more* in the direction normal to the surface than in any other direction; this condition is likely easier to verify in practice. In any sense, the example shown in Figure 3.3b is certainly not normally hyperbolic since there is no expansion of the line element in the direction normal to the material surface (*x*-axis).

Inevitably in most practical applications, stretching along a hyperbolic material surface will have tangential and normal components, placing its behavior somewhere between the ends of the spectrum presented in Figure 3.3, with the additional complication that the stretching characteristics can change along the surface in both space and time. Refinements to our views of hyperbolicity offer alternatives for us to refine how we ultimately define a maximally stretching surface as an LCS. At this point, only a few works have distinguished different categories of maximal stretching surfaces in practical applications. Preliminary results indicate that most structures are normally hyperbolic [46]; however, other types of structures are certainly possible [47]. More experience is needed at this time to understand the implications

---

7) In the saddle flow, one may notice that the growth is exponential in time, whereas the growth in the shear flow is linear in time. From these examples, one might expect that for sufficient integration time, normally hyperbolic structures become more pronounced than shear structures when plotting FTLE. While this may tend to happen, it is not always the case. Furthermore, distinction of "exponential growth" for finite-time, nonperiodic systems is not straightforward; for example, the shear flow has nonzero FTLE – more generally, any type of growth in finite time can be bounded by an exponent.

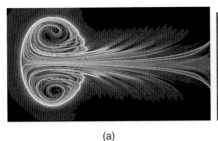
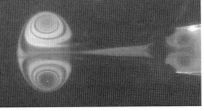

(a)   (b)

Figure 3.4 Comparison of backward time FTLE and attracting LCS (panel a) with a florescence visualization (panel b) by Paul S. Krueger and John O. Dabiri. The PIV data used to compute FTLE and the florescence visualization come from similar experimental setups, but different experiments and different times during each experiment. Nonetheless, structures revealed in both fields are similar, as florescence aligns with attracting LCS. The role of repelling LCS is less obvious from the flow visualization.

of such distinctions in practical flows. This could perhaps lead to deeper understandings on the nature of fluid, and help refine how we approach studying it.

We note that any notion of hyperbolicity must be associated to a specific time interval, exemplifying one of the main idiosyncrasies of finite-time analysis. Thus, our view on the stability of a surface in general depends on our horizon of knowledge. For example, perturbations to a surface could be marked by mostly tangential stretching during one time interval and switch to primarily normal expansion using a slightly longer horizon, or a surface could be marked by expansion using one integration time and then attraction using another. Likewise, a ridge in the FTLE field may only appear as a ridge using a set range of integration times.

Finally, we point out that attracting LCS will generally be distinguishable from advection of material points since fluid is attracted to and along these surfaces. This point connects to arguably the most intuitive method of understanding flow – the visualization of some visual marker that follows the fluid, "tracers." The utility of flow visualization is not to track individual particles but to use tracers to reveal coherent features that reveal how the flow is organized. We now know that such coherent patterns commonly observed are manifestations of attracting LCS (cf. Figure 3.4). However, we point out that repelling LCS are essentially hidden from flow visualization, even though these structures play a fundamental role in transport. Therefore, the computation of attracting and repelling LCS both broadens (by revealing a new class of hidden coherent structures) and deepens (by making the organizing surfaces precise) our understanding of coherent structures in fluid flow.

3.3.4 Objectivity

A desirable property of the global approach is that it is independent of coordinate frame, that is, objective. This cannot be said of most common flow diagnostics.

A quantity is called objective if it remains invariant under translations and rotations, that is, coordinate transformations $x \mapsto y$ of the form

$$y = Q(t) x(x_0, t_0, t) + b(t) \tag{3.12}$$

where $Q(t)$ is a time-dependent rotation matrix and $b(t)$ represents a time-dependent translation. As before, let

$$\xi_t = x(x_0 + \xi_0, t_0, t) - x(x_0, t_0, t) \tag{3.13}$$

Applying a general transformation of coordinates according to Eqs. (3.12) and (3.13) becomes

$$\zeta_t \doteq Q(t) x(x_0 + \xi_0, t_0, t) + b(t) - Q(t) x(x_0, t_0, t) - b(t) \tag{3.14}$$

$$\zeta_t = Q(t) \xi_t \tag{3.15}$$

Note, $Q(t)$ is an isometry, so $|\zeta_t| = |\xi_t|$. Therefore, *relative* stretching measures such as Λ remain invariant under coordinate transformation.

3.4
Computational Strategy

LCS are effectively revealed by computation of the FTLE field. The basic strategy for computing the FTLE field is straightforward. Essentially, a grid of material points is advected and the deformation of the grid is used to compute the distribution of FTLE. Notably, to compute the FTLE *field*, we need trajectory information over a *grid* of points. Out of necessity, trajectory information is usually derived from postprocessing velocity field data obtained by computational fluid dynamics or empirical measurement. While computationally intensive, this approach is highly flexible, allowing for variations in the distribution and resolution of trajectory information, which can ultimately influence the accuracy and performance of LCS identification. The use of trajectory information independent of velocity field information is possible, and indeed desirable since it avoids introducing modeling and computational errors, but rarely a viable option.

The standard algorithm starts with the initialization of a grid of material points at time t_0. For simplicity, we assume these points form a structured grid $x_{ijk}(t_0)$. This simplifies gradient computation and does not require *a priori* knowledge of the flow topology, which makes the discussion most broadly applicable. The initial locations of these points, as opposed to the final locations, represent the locations at which FTLE will be computed – the FTLE grid. The material points are integrated by the velocity field for some finite time interval, $T = t - t_0$,

$$x_{ijk}(t) = x_{ijk}(t_0) + \int_{t_0}^{t} u(x_{ijk}(\tau), \tau) d\tau \tag{3.16}$$

using numerical integration. Note, for measures such as FSLE or HT, the integration time, T, of each trajectory depends on the stretching characteristics, whereas measures such as the FTLE, RD, or FS nominally use a fixed horizon for each material point. Such integration requires the velocity field to be interpolated in space and time. Higher-order integration and interpolation schemes ensure greater accuracy and smoothness of the computed results, respectively. One may assume that since we have set out on a numerically daunting task of measuring exponential growth, and computing quantities based on first and higher order derivatives of the flow, that the integration and interpolation schemes are critical. Even though precise trajectory computation is difficult in chaotic flow, LCS identification is typically reliable (cf. Section 3.5). Notably, integration and interpolation errors do not typically produce spurious structures, nor will they cause influential LCS to be undetected. Of course variations in identified LCS location, smoothness, and so on will depend on the numerical method, but variations are typically small if care is taken to ensure the integration and interpolation schemes used are reasonable in the light of the flow conditions. Efficient integration on unstructured grids was described in [48].

Once the final positions, $x_{ijk}(t)$, are computed, the deformation gradient can be evaluated at each point in the structured grid $x_{ijk}(t_0)$ by finite differencing, for example, using central differencing

$$\nabla F_{t_0}^t \big|_{(x_{ijk}(t_0), y_{ijk}(t_0), z_{ijk}(t_0))} \approx \begin{bmatrix} \frac{x_{(i+1)jk}(t) - x_{(i-1)jk}(t)}{x_{(i+1)jk}(t_0) - x_{(i-1)jk}(t_0)} & \frac{x_{i(j+1)k}(t) - x_{i(j-1)k}(t)}{y_{i(j+1)k}(t_0) - y_{i(j-1)k}(t_0)} & \frac{x_{ij(k+1)}(t) - x_{ij(k-1)}(t)}{z_{ij(k+1)}(t_0) - z_{ij(k-1)}(t_0)} \\ \frac{y_{(i+1)jk}(t) - y_{(i-1)jk}(t)}{x_{(i+1)jk}(t_0) - x_{(i-1)jk}(t_0)} & \frac{y_{i(j+1)k}(t) - y_{i(j-1)k}(t)}{y_{i(j+1)k}(t_0) - y_{i(j-1)k}(t_0)} & \frac{y_{ij(k+1)}(t) - y_{ij(k-1)}(t)}{z_{ij(k+1)}(t_0) - z_{ij(k-1)}(t_0)} \\ \frac{z_{(i+1)jk}(t) - z_{(i-1)jk}(t)}{x_{(i+1)jk}(t_0) - x_{(i-1)jk}(t_0)} & \frac{z_{i(j+1)k}(t) - z_{i(j-1)k}(t)}{y_{i(j+1)k}(t_0) - y_{i(j-1)k}(t_0)} & \frac{z_{ij(k+1)}(t) - z_{ij(k-1)}(t)}{z_{ij(k+1)}(t_0) - z_{ij(k-1)}(t_0)} \end{bmatrix}.$$

(3.17)

Once the deformation gradient is computed, straightforward evaluation of its largest singular value, and subsequently the FTLE (or similar), at each location $(x_{ijk}(t_0), y_{ijk}(t_0), z_{ijk}(t_0))$ is possible. This procedure provides the FTLE field and thus reveals LCS at the time instant t_0. For unsteady flow, the FTLE field and locations of LCS generally change over time and the above procedure can be repeated for a range of times t_0 to provide a time series of FTLE fields, and consequently, a time history of the LCS movements. Note that to compute a time series of FTLE fields, a new grid of material points is "released" at a series of release times t_0 and integrated from t_0 to $t_0 + T$.

A few notes are in order. First, we recall that the linearized flow map, that is, deformation gradient, can be derived from solving the linearized dynamical system

$$\frac{d}{dt} \nabla F_{t_0}^t(x_0) = \nabla u(x(t), t) \cdot \nabla F_{t_0}^t(x_0) \tag{3.18}$$

ergo perturbations can be in theory integrated by the linearized flow. Nonetheless, we typically integrate the nonlinear flow and then use finite differencing to approximate the deformation gradient. In practice, the two methods are not the same due to the finite differences involved. Second, the computation of trajectories is highly parallel since particles are assumed to not interact with each other or the flow. Therefore, one can readily exploit parallel architectures for trajectory computation by partitioning in space (or time when computing a time series of FTLE fields), see, for example, [49, 50]. Third, in the computation of a time series of FTLE fields, the integration interval for a particular release will typically overlap with the integration time interval of later, or previously, released grids. Therefore, it would be advantageous if each release could "use the same flow map information." This is typically not possible since material points in each release will not occupy the same position at the same time; however, Brunton and Rowley [51] demonstrate that under suitable approximation, an efficient flow map interpolation strategy can be devised. Finally, several variations of the above procedure are possible and in many cases desirable; the steps outlined here represent the rudimentary components.

3.4.1
Grid-Based Computation

Here we point out a few practical comments regarding the use of the FTLE grid. First, the computation of the deformation gradient described above is equivalent to considering the stretching of three orthogonal line elements, in the x-, y-, and z-directions, with each column representing the stretching from each line element. From Eq. (3.3) and the discussion that followed, we might expect that stretching of each element becomes similar. Indeed, the FTLE could be approximated at each point by considering the stretching of a single line element. This, in fact, has been done, see, for example, [25, 26, 29]. However, the evaluation of the full-deformation gradient does not incur substantial computational expense, and can provide more accurate results and additional information that could be useful for LCS identification (cf. Section 3.2). Furthermore, several related stretching measures can be derived from the deformation gradient, and using elements that span all linearly independent directions, we are more likely to "straddle" all LCS (cf. discussion below), than when using a single line element at each grid point.

In Section 3.1, we showed that the FTLE defines the average logarithmic expansion rate about a trajectory for *infinitesimal* perturbations. Likewise, the entries in Eq. (3.17) should be defined from infinitesimal elements to determine the deformation gradient. In the computation of FTLE, the neighbors in the FTLE grid itself are typically used for finite differencing. Using this strategy, the grid should, in theory, be highly resolved. For most applications, the resolution needed would, in fact, be cost prohibitive. The exponential growth we seek to reveal quickly leads to "overflow" as the integration proceeds (cf. Winkler [28], Section 5.4 for estimates) and we would need to integrate more material points than is often practically possible to remain under theoretical limits. Alternatively, at each grid point, one can consider the advection of local perturbations that are much smaller in magnitude than the FTLE grid spacing. In theory, these perturbations could be made small enough so that arbitrary accuracy in the FTLE computation is achieved. Similarly, one can rescale the perturbation as it grows too large – a traditional approach in computing Lyapunov exponents. However, the actual value of the FTLE we compute is not our primary motivation – the computation of LCS is our primary motivation; thankfully so, since in most cases, computing FTLE by finite differencing of the FTLE mesh does not accurately capture the theoretical FTLE values (cf. Section 2.3 of the work by Tabor and Klapper [52], and references therein, for discussions on the stretching of finite elements).

It is in fact the "straddling" of these manifolds that appears more important (which is likely why "less rigorous" stretching measures often reveal LCS); for example, refer back to the FTLE field shown in Figure 3.2. Notice that the value of FTLE can quickly decrease away from the LCS. Therefore, even if we perfectly computed the FTLE values over a coarse grid using infinitesimal perturbations, we may not likely see any ridges in the FTLE field since generally we would not expect the grid points to lie on, or sufficiently close to, the LCS. However, by differencing the computational grid as outlined above, LCS that lie between grid points will be detected, even for relatively coarse meshes. Consequently, if one is interested in knowing the approximate location of the LCS, then a coarse grid can be used to obtain the approximate LCS location, and then the FTLE grid can be adaptively refined near LCS to iteratively improve the location estimates. We originally discussed these ideas in [53]. Sadlo *et al.* [54, 55] implemented a grid adaptation method that targeted FTLE ridges; however, they used a more targeted approach that, instead of refining results from a coarse FTLE grid, used other *a priori* knowledge of the flow as a starting point for refinement. Similarly, Garth *et al.* [56] implemented an approach similar in spirit where grid refinement was based on accuracy of flow map determination, rather than FTLE ridges (note, though, the flow map is, in a sense, most sensitive near LCS).

If the grid spacing is too large, then neighboring grid points can span multiple LCS, or portions of the same LCS, and results become uncorrelated with any particular LCS. More generally, as the grid spacing increases, the FTLE field becomes uncorrelated with the actual flow topology. As grid spacing decreases, LCS detection is improved; however, computational cost grows exponentially. There is a common misconception that the FTLE grid need not be more resolved than the velocity grid used to drive the computations. However, there should be a distinction between Lagrangian and Eulerian information. Certainly, two points starting inside the same element of the velocity mesh are not expected to have the same dynamics. Indeed, for example, it is certainly possible for more than one LCS to lie between velocity grid points. Generally, the FTLE mesh is more highly resolved than the velocity mesh.

3.4.2
Integration Time

The Lagrangian perspective is of central importance to the study of LCS, which bases our analysis on the behavior of the fluid over time. When computing FTLE, or similar, an integration time should be chosen that is long enough, so the dominant flow features have a chance to emerge, but not so long that the final positions used to compute the deformation gradient are uncorrelated to the flow features we are trying to expose back at the initial time the material points were released. The choice of integration time used to compute FTLE is often *ad hoc*, and in many applications, this is little concern. In general, LCS can shrink, grow, appear, and disappear with changes in integration time. Typically, LCS are robust and long-lived (compared to Eulerian time scales), and thus variations of the FTLE topography with integration time are gradual. Specifically, the following generalizations often hold: (1) As integration time is initially increased, FTLE distribution tends to sharpen along LCS and the exposed length of LCS tends to grow; the growth often results from the FTLE computation tracking how the influence of hyperbolic hubs extend out into the domain. (2) The exposed location of a particular LCS in the FTLE field is typically not sensitive to changes in integration time (except perhaps for relatively short integration times when LCS are not well-defined). That is, holding t_0 fixed in the FTLE computation, the identified location of LCS should remain fixed if the LCS is a material surface.

Consider again Figure 3.3a. The particles appear to exponentially diverge near the hyperbolic fixed point. The further away the particles are from the fixed point along the stable manifold, the longer it takes for them to reach, and be significantly stretched apart by, that hub. Considering LCS as a stable manifold of a hyperbolic trajectory, stretching at a particular section of the LCS may not become pronounced until one considers a sufficiently long time horizon. For example, Figure 3.5 shows the variation of the FTLE with integration time for the double-gyre system considered in [32] given by the stream function

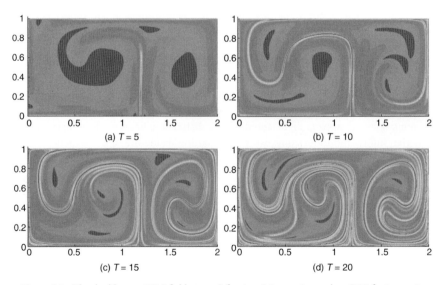

Figure 3.5 The double-gyre FTLE field at $t = 0$ for $A = 0.1$, $\omega = 2\pi$, and $= 0.25$ for increasing integration times T.

$$\gamma(x, y, t) = A \sin(\pi f(x, t)) \sin(\pi y) \tag{3.19}$$

where

$$\begin{aligned} f(x,t) &= a(t)x^2 + b(t)x \\ a(t) &= \varepsilon \sin(\omega t) \\ b(t) &= 1 - 2\varepsilon \sin(\omega t) \end{aligned} \tag{3.20}$$

over the domain $[0, 2] \times [0, 1]$. There is a stable manifold attached to the bottom boundary and as the integration time is increased, more of the manifold is revealed in the FTLE plot.

One may notice in Figure 3.5d that a segment of the LCS close to the hyperbolic trajectory has become faint, which is not a rendering artifact. This may seem paradoxical; however, material points that were released from this location have started to reach a condition where their final locations are "uncorrelated" with the local flow topology. A higher grid resolution is needed for the averaged growth rate over this integration time to be representative of the local flow topology at this location. Note that this occurs mainly for systems for which material points are likely to revisit the same region; for example, one does not readily observe this phenomenon for the vortex example (Figure 3.2), which is why we choose instead to show the variation of T for the double gyre flow.

Structured approaches to determining adequate integration times have been proposed. Winkler [28] used ideas from Boffetta *et al.* [57] to predict appropriate integration times for various examples in computing RD to restrict stretching to the "exponential regime," thereby limiting the influence of longer term diffusive behavior. In the double-gyre flow, the hyperbolic trajectory on the bottom boundary persists for all time. In most practical applications, hyperbolic trajectories may exist only over finite time intervals, and similarly for their stable and unstable manifolds. Therefore, as the integration time for the FTLE becomes large, it is possible that LCS existing over various subsets of time begin to "cloud" the FTLE field, and subsequently our interpretation of the flow. Lekien and Leonard [58] provided a structured approach to filter out appropriate time scales from the dynamics to determine an appropriate integration time. As mentioned toward the end of Section 3.2, the direction of dominant expansion/contraction converges to the normal direction for a normally hyperbolic LCS; in fact this is a necessary condition [22]. Therefore, the time needed for this alignment to converge (up to some ε) can be taken as a lower bound for the integration time needed to identify normally hyperbolic LCS.

From the practical view, the velocity data is only known over a finite interval of time and, for nonperiodic flow, the integration of trajectories used to compute FTLE cannot proceed beyond the timespan of the data set. Furthermore, unless all boundaries of the spatial domain of the data set are invariant, material points used to compute the FTLE field are often flushed from the domain before the desired integration time is reached, unless the open boundaries are sufficiently far from the region one is computing FTLE. Quite frequently, practical limits are imposed on the integration time length before dynamics-based limits become relevant, especially when computation and measurement of the velocity data is performed with Eulerian features in mind. To summarize, integration time is usually determined by grid spacing (computational cost), time scale of the dynamics, and/or availability of data.

In the FTLE computation, material points that leave the domain before the desired integration time can be treated in a number of ways. Often it is not possible to extrapolate the velocity field outside the bounds of the data set without a high degree of arbitrariness. For material points that exit early, it is common to compute the FTLE for that point, and due to the finite-differencing strategy (cf. Eq. (3.17)), at neighboring points in the FTLE grid.[8] A drawback is that various parts of the domain can end up using different integration times to compute their FTLE value, and notably, the stretching rate (logarithmic or otherwise) is typically not constant over a trajectory in finite time. Therefore, this strategy can result in spurious structures in the FTLE field, for example, marking boundaries to portions of fluid that have exited the domain. Also, in the author's experience, applications where material points are likely to exit during FTLE computation, the $1/|T|$ scaling in the definition of the FTLE can often amplify the FTLE values of points that leave within a small integration time and it is often undesirable to include this scaling. Conversely, in applications where points do not leave the domain, the $1/|T|$ scaling is unnecessary, as it only shifts all values uniformly.

Along with spurious LCS that can result from open domains, limited data sets can lead to dominant coherent structures failing to emerge before particles exit the domain. In such cases extrapolation is often desirable, even if arbitrary. For example, in applications of studying LCS in heart flow problems, we have often extended the velocity field information by creating linear fields outside of the domain of data, for example, to represent blood pumped out of the heart. Tang *et al.* [59] developed a similar approach of extending atmospheric flow data using linear velocity fields to study clear air turbulence over the Hong Kong airport, and introduced a smoothing to ensure C^1 continuity of the fields. The advantage of linear fields is that they maintain a homogeneous stretching rate outside the domain. Nonetheless, LCS revealed in the original domain of data can be subject to the extrapolated velocity field in nonlinear ways. For example, for neighboring points in the FTLE grid, one point can be advected outside of the domain into the linear field, whereas the other point may remain inside the domain for a significantly longer time. The subsequent stretching is due to a combination of the linear flow with the original nonlinear flow, making the combination nonlinear and the stretching rate nonconstant. This is more of a concern in applications with open and closed boundaries, as opposed to applications with strictly open boundaries. Also, the linear field outside the data domain may look nothing like the actual velocity field outside the data domain. However, the use of linear fields tends to minimize the introduction of additional arbitrary stretching

---

8) When implemented as such, this excludes "cases when neighboring trajectories all leave the domain earlier than the trajectory considered" [59], since the FTLE would already have been computed for that trajectory. Note that this approach does not stop integration prematurely (as that would cascade through the grid), it just computes FTLE prematurely.

since it holds stretching rates constant once elements leave the domain, or as put by Tang *et al.* [59] tends to "lock-in and enhance" existing FTLE ridges.

As mentioned previously, positive integration times reveal repelling LCS in the FTLE field and negative integration times reveal attracting LCS in the FTLE field. Expansion is nominally easier to compute than contraction; therefore, we reverse time to effectively measure contraction as expansion. In theory however, when we advect a grid of material points, both types of information are encoded. Indeed, Haller and Sapsis [60] recently showed that when the integration time is positive, troughs of the *smallest* FTLE field reveal attracting LCS (or equivalently ridges of the negative of the smallest FTLE). The advantage of this approach is that only a single numerical run is needed, instead of separate numerical runs in forward and backward time. Note, however, that the smallest FTLE field is plotted at the *final* locations of the material points. Therefore, the resulting field is defined over a highly distorted grid, which often must be interpolated to a regular grid for analysis.

3.4.3
LCS Extraction

As presented here, the computation of FTLE fields is often seen as a means to an end; ultimately, we desire to locate LCS to better understand the flow topology. FTLE provides a measure that enables relatively easy and robust determination of LCS location *and* strength. We use FTLE precisely because it highlights these structures; therefore, one could argue that the more we can highlight, or filter out, LCS from the background FTLE field the better. Initial applications of the framework focused on demonstrating the existence and implications of these structures, which indeed was critical to this method gaining acceptance and wider use. Now that the subject has begun to mature, LCS are becoming part of, or have the potential to become part of, broader computational objectives, for example, decision making, design, transport calculations, and optimization. In this light, it is often necessary to have the ability to effectively parameterize or extract LCS from FTLE plots. Finally, LCS filtering or extraction is necessary for effectively visualizing these structures in 3D. Not surprisingly how to best extract LCS is closely related to determining what is the best way to allocate the FTLE computational costs to target LCS.

For 2D systems, we often *visually* filter out LCS by mapping colors (which could include intensities of a particular color) to FTLE values. In such cases, ridges are visually assessed from the color (e.g., Figures 3.2 and 3.5) or intensity (e.g., Figures 3.4 and 3.6) mapping. In general, when dealing with FTLE values themselves, we are temped to associate LCS as sets of globally high FTLE (e.g., threshold out low FTLE values); however, FTLE ridges are defined in terms of spatial derivatives of the FTLE field (Section 3.2) not the actual FTLE values per se. In 3D, visual identification of LCS from unfiltered color or intensity mapping is not possible. One must either extract the LCS surfaces, or use volume rendering to visualize the LCS in the FTLE plots (or reduce to 2D by considering sections). In regards to volume rendering, the primary goal is essentially to skip as much of the volume as possible while retaining the subvolumes containing the LCS. In the most basic sense, volume rendering can be achieved by mapping the FTLE to color and opacity scales, whereby FTLE becomes transparent as it decreases (thresholding can be seen as a naive implementation). This works well for visualizing the most well-defined LCS. At least in 2D, opacity mappings can also be helpful when showing multiple fields (e.g., superimposing forward and backward time FTLE fields); for example, see work of Garth *et al.* [56] for a systematic procedure that couples FTLE plots with a "texture-based" representation [61] of the flow and leverages GPU to accelerate FTLE computation.

We have previously discussed (e.g., [32, 48, 53]) how LCS can be extracted based on the definition in Section 3.2. Using previous notation, let $\mathbf{v}_{min}$ be the unit eigenvector associated μ_{min}, which must be normal to the LCS. Along an LCS $\nabla \Lambda \cdot \mathbf{v}_{min} = 0$ and therefore potential LCS can be extracted from zero level sets of the field $\alpha = \nabla \Lambda \cdot \mathbf{v}_{min}$. There is a sign ambiguity with computing $\mathbf{v}_{min}$ and therefore the sign must be chosen in a locally consistent manner – this is critical for proper ridge extraction. This method has been applied in various applications [48, 53, 62], often when specific LCS were used for transport rate computations, to various degrees of success. A difficulty with this approach is that smooth second derivatives are notoriously difficult to compute from discrete data, and the computation of $\nabla^2 \Lambda$ is no exception. Generally, this approach distinguishes *all* ridges of FTLE, notably many of which our eyes would pass over in the FTLE field. Therefore, one looses the relative strength information encoded in the FTLE plot and significant contextual information. Hence, LCS extraction is perhaps more critical when driving additional computations where an exact parameterization is needed than for visual appeal. It is usually desirable to filter spurious ridges or weak LCS by checking additional conditions, such as setting a minimum threshold for Λ or maximum threshold for μ_{min}; the first indicating strength of hyperbolicity and the second how well-defined the ridge is, which can be tied to Lagrangian [32, 33] and robustness [22] properties. The application of ridge extraction to the mechanically generated vortex is show in Figure 3.6. Unfortunately, this filtering can cause portions of otherwise well-defined LCS to be skipped where the criteria locally fail (presumably due to computational anomalies) as also seen in the figure. A similar ridge extraction approach was adopted by Sadlo *et al.* [54, 55] who also noted troubles with the practical application of this method in noisy data sets and offered some insights on addressing these problems.

Mathur *et al.* [46] introduced a clever way to distinguish FTLE ridges. Essentially, once the FTLE field is computed, material points are seeded near LCS and advected by the FTLE gradient instead of the flow. In this way, the gradient of the FTLE field $\nabla \Lambda$ pushes the points up onto the ridge. Certain criteria check once each point approximately reaches the top of the ridge and the computation for the point is stopped. In other work, Lipinski and Mohseni [63] proposed a ridge-tracking algorithm for efficient LCS computation in 2D. The basic idea of this method is to target the FTLE computation along LCS, and in doing so provide tight estimates of the LCS locations. The approach starts by computing FTLE along lines that intersect the domain. Local maxima of FTLE along these lines are assumed to correspond to intersections with LCS. These locations, and estimates of the local orientation of the

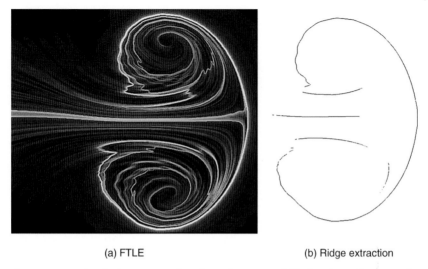

(a) FTLE (b) Ridge extraction

Figure 3.6 Results of ridge extraction from forward time FTLE field of mechanically generated vortex ring. Ridges with $\Lambda < 70\%$ of maximum, and $\mu_{min} > -4800$ are not shown.

LCS, are used to allocate FTLE mesh points only in the directions of the LCS. FTLE is then computed at the newly seeded points to further the location and orientation estimate, and the process continues sequentially to grow ("track") the location of the LCS in both spatial directions.

We note that in most cases, the extraction of LCS from FTLE fields has been directed at 2D problems, or in 3D problems by considering 2D sections. However, the need for extraction is arguably most compelling in 3D. We also note that since LCS are material surfaces, it has been observed that if one is able to extract LCS at one time instant, the time evolution of the LCS can be obtained by advecting it as a material surface (e.g., this is leveraged in [55, 63, 64]). However, because LCS can shrink, grow, appear, and disappear, it is unclear how broadly useful this is in practice. Greatest utility may be for attracting LCS [64]. That is, high FTLE values indicate where particles are *most* sensitive to changes in initial conditions; therefore, errors in the position estimate for repelling LCS will quickly grow, whereas for attracting LCS they will shrink. More notably though, because of the typical normal expansion about repelling LCS, fluid along the surface tends to compress, decreasing the scope of LCS revealed over time. Presumably, one may consider the advection of a repelling LCS *backward* in time to efficiently display past locations of repelling LCS, or advect attracting LCS as material surfaces forward in time to display future locations. However, in real-time applications, future positions of the repelling LCS are most sought after since these structures require future knowledge to obtain. That is, if data is known up to the real time t and an integration time of $|T|$ is used for FTLE computation, repelling LCS can only be computed up to at past time $t_0 = t-|T|$, using a traditional approach. Hence, for a realtime estimate, the material surface must be advected forward in time, subject to the prior mentioned difficulties.

3.5
Robustness

The computation of LCS is typically derived from trajectory information obtained by processing Eulerian velocity data. The velocity data often contains inherent modeling or measurement errors. Furthermore, since the velocity data is discrete, the integration process must fill the gaps in space and time with an interpolated field, which introduces additional deviations from the true field. Even if these errors are small, they can accumulate during integration and lead to significant errors in the final position estimates used to compute the deformation gradient. Thus, one may wonder if the location, or even existence, of a particular LCS holds for the true system, or just the "model" system? This is a particularly difficult question to answer since one typically does not know the exact nature of the errors involved. One can approach studying this problem in a number of ways. Theoretically, one can choose a certain class of errors and derive rigorous estimates for the robustness of LCS. Empirically, one can systematically introduce known differences between the true and model fields and quantify changes in the results, or compare results from realizations with unknown but distinct errors. Finally, one can compare results of LCS computations with independent measurements indicating LCS location from the true system. All of these have been done. In general, it has been shown that LCS are surprisingly robust considering the integrated error influence.

In Ref. [31], Haller showed that if the model velocity field is close in a time-weighted norm to the true velocity field, then a hyperbolic material line of the model system implies a hyperbolic material line in the true system. The weighted norm is such that it permits relatively large errors as long as they are local in time – those that do not accumulate to large errors when integrated over time. Haller showed, as one might otherwise assume, that strongly repelling/attracting and/or long-lived LCS are the most robust, which more recently [22] was translated into conditions on FTLE ridges, which essentially imply that the ridge should be well-defined. Olcay et al. [65] investigated how velocity field resolution and random errors affect LCS. In their approach, they degraded a vortical flow field by locally smoothing the flow and sampling it on spatially sparser grids, subsampling the data in time, and adding Gaussian noise. They found that mean LCS locations were robust to these variations in the velocity data, but locally the LCS could significantly deviate from its baseline location. In all cases, the existence of the LCS was robust. Shadden et al. [62] compared LCS computed from radar measurements of ocean surface currents with GPS-tracked surface buoy data. They showed that even though the highly noisy radar data led to errors in trajectory computation that approach 20–30 km over the integration interval used to compute FTLE, the error in location of the LCS appeared to be on the order of only a few kilometers. More recently, Harrison and Glatzmaiera [66], in studying the California Current system, compared LCS computed from a mesoscale eddy field generated from Fourier filtering satellite altimetry observations and from the full observational satellite altimetry. They showed that the LCS found were relatively insensitive to both sparse spatial and temporal resolution, and to the velocity filtering.

3.6
Applications

The computation of LCS has proven to be effective in wide range of fluid mechanics problems. Applications of this method have ranged from microfluidic to geophysical and from laminar to turbulent. In addition to this wide scope, the recent growth in popularity of this approach is also likely due in large part to this method being relatively straightforward to implement and hence highly accessible. Here, we briefly list some representative publications describing the computation of LCS to study practical fluid transport problems. In many of these publications, unique perspectives on LCS computations, as well as their relevance to understanding transport processes in the respective application, are presented.

Transport in the ocean and atmosphere has wide-ranging importance. Historically, the computation of stable and unstable manifolds has seen preference in this area, and not surprising the earliest computations of LCS using FTLE fields focused on these application areas [23, 25, 26, 29]. More recent studies of atmospheric mixing include the works of Refs. [47, 67, 68] and studies of the polar vortices include the works of Refs. [69–71]. In regards to ocean mixing, publications include Refs. [62, 66, 72–79]. Beron-Vera, Olascoaga *et al.* [80–86] have provided a number of fruitful applications of this method to marine science. Tang and Peacock explored the implications of the approach for locating internal ocean waves [87]; several groups have used this approach to studying transport associated with hurricanes/cyclones [8, 14, 88, 89] and to study transport of ocean [90] and atmospheric [91] contaminants.

The use of LCS has also been used extensively to study vortex rings [42, 92–94], as well as hairpin vortices in turbulent channel flow [95], and vortex shedding [96–99]. In aeronautics, LCS has been used to study clear air turbulence [100] and flow separation [32, 101]. In the realm of biofluids, LCS applications include studies of respiration [102], animal swimming [5, 103–110], and cardiovascular fluid mechanics [48, 111–113]. Finally, several studies have used LCS to understand the mechanics of fluid mixing devices [46, 53, 114].

3.7
Conclusions

LCS are core surfaces that organize fluid transport. LCS computation arose as a means to compute effective stable and unstable manifolds of hyperbolic trajectories in temporally aperiodic systems, which was motivated by previous discoveries that such manifolds organized transport in temporally periodic flows. Indeed, in periodic and aperiodic systems alike, these structures often govern the stretching, folding, and alignment mechanisms that control kinematic mixing. As the application of this framework branched out to increasingly complex problems, LCS have become more broadly associated with distinguished material surfaces. Notably, while the entirety of a fluid is composed of a continuum of material surfaces, those with locally the strongest stability (attracting or repelling) are often influential in dominating advection patterns. The computation of LCS has enabled the ability to better perceive and appreciate this influence. To date, LCS are effectively identified by plotting Lagrangian-based measures of hyperbolicity, such as the FTLE. We have shown how such computations proceed in practice, and a number of theoretical and practical concerns associated with this approach. Wide-ranging applications have shown the utility of LCS toward enabling deeper understanding of transport-related phenomena over a wide range of spatial scales and temporal variability. Continued theoretical, computational, and application-oriented developments in this regard will hopefully further enable greater comprehension of the nature of unsteady fluid transport and its implications.

Acknowledgments

This chapter has benefited from the keen insight of George Haller, who provided valuable recommendations, and more generally helped pave the way for LCS computation.

References

1 Wiggins, S. (1994) *Normally Hyperbolic Invariant Manifolds in Dynamical Systems*, Springer, New York.
2 Ottino, J.M. (1989) The kinematics of mixing: Stretching, chaos, and transport, in *Cambridge Texts in Applied Mathematics*, Cambridge University Press, Cambridge; New York, J.M. Ottino. ill.; 24 cm.
3 Wiggins, S. (1992) *Chaotic Transport in Dynamical Systems*, Springer, New York.
4 Aref, H. (2002) The development of chaotic advection. *Physics of Fluids*, **14** (4), 1315–1325.
5 Peng, J. and Dabiri, J.O. (2009) Transport of inertial particles by Lagrangian coherent structures: Application to predator-prey interaction in jellyfish feeding. *Journal of Fluid Mechanics*, **623**, 75–84.
6 Haller, G. and Sapsis, T. (2008) Where do inertial particles go in fluid flows? *Physica D*, **237**, 573–583.
7 Sapsis, T. and Haller, G. (2010) Clustering criterion for inertial particles in two-dimensional time-periodic and three-dimensional steady flows. *CHAOS*, **20** (1), 017515.
8 Sapsis, T. and Haller, G. (2009) Inertial particle dynamics in a hurricane. *Journal of the Atmospheric Sciences*, **66** (8), 2481–2492.
9 Tallapragada, P. and Ross, S.D. (2008) Particle segregation by Stokes number for small neutrally buoyant spheres in a fluid. *Physical Review E*, **78** (3), Part 2).
10 Miller, P.D., Jones, C.K.R.T., Rogerson, A.M., and Pratt, L.J. (1997) Quantifying transport in numerically generated velocity fields. *Physica D*, **110** (1–2), 105–122.
11 Haller, G. and Poje, A.C. (1998) Finite time transport in aperiodic flows. *Physica D*, **119** (3–4), 352–380.
12 Ide, K., Small, D., and Wiggins, S. (2002) Distinguished hyperbolic trajectories in time-dependent fluid flows: Analytical and computational approach for velocity fields defined as data sets. *Nonlinear Processes in Geophysics*, **9**, 237–263.
13 Haller, G. and Yuan, G. (2000) Lagrangian coherent structures and mixing in two-dimensional turbulence. *Physica D*, **147** (3–4), 352–370.
14 du Toit, P. and Marsden, J.E. (2010) Horseshoes in hurricanes. *Journal of*

Fixed Point Theory and Applications, **7**, 351–384.

15 Romkedar, V., Leonard, A., and Wiggins, S. (1990) An analytical study of transport, mixing and chaos in an unsteady vortical flow. *Journal of Fluid Mechanics*, **214**, 347–394.

16 Romkedar, V. (1990) Transport rates of a class of 2-dimensional maps and flows. *Physica D*, **43** (2–3), 229–268.

17 Fenichel, N. (1971) Persistence and smoothness of invariant manifolds for flows. *Indiana University Mathematics Journal*, **21**, 193–225.

18 Mancho, A.M., Small, D., Wiggins, S., and Ide, K. (2003) Computation of stable and unstable manifolds of hyperbolic trajectories in two-dimensional, aperiodically time-dependent vector fields. *Physica D*, **182** (3–4), 188–222.

19 Duc, L.H. and Siegmund, S. (2008) Hyperbolicity and invariant manifolds for planar nonautonomous systems on finite time intervals. *International Journal of Bifurcation and Chaos*, **18** (3), 641–674.

20 Froyland, G., Padberg, K., England, M.H., and Treguier, A.M. (2007) Detection of coherent oceanic structures via transfer operators. *Physical Review Letters*, **98**, 224503.

21 Froyland, G. and Padberg, K. (2009) Almost invariant sets and invariant manifold-connecting probabilistic and geometric descriptions of coherent structures in flows. *Physica D*, **238**, 1507–1523.

22 Haller, G. (2011) A variational theory of hyperbolic Lagrangian coherent structures. *Physica D*, **240**, 574–598.

23 Bowman, K. (1999) Manifold geometry and mixing in observed atmospheric flows, http://geotest.tamu.edu/userfiles/213/manifold_geometry.pdf.

24 Jones, C.K.R.T. and Winkler, S. (2002) Invariant manifolds and lagrangian dynamics in the ocean and atmosphere, in *Handbook of Dynamical Systems*, vol. 2 (ed. B. Fiedler), Elsevier, Amsterdam, pp. 55–92.

25 Pierrehumbert, R.T. (1991) Large-scale horizontal mixing in planetary-atmospheres. *Physics of Fluids A*, **3** (5), 1250–1260.

26 Pierrehumbert, R.T. and Yang, H. (1993) Global chaotic mixing on isentropic surfaces. *Journal of the Atmospheric Sciences*, **50** (15), 2462–2480.

27 Doerner, R., Hubinger, B., Martienssen, W., Grossmann, S., and Thomae, S. (1999) Stable manifolds and predictability of dynamical systems. *Chaos Solitons & Fractals*, **10** (11), 1759–1782.

28 Winkler, S. (2001) Lagrangian dynamics in geophysical fluid flows. Ph.D., Brown University.

29 von Hardenberg, J., Fraedrich, K., Lunkeit, F., and Provenzale, A. (2000) Transient chaotic mixing during a baroclinic life cycle. *CHAOS*, **10** (1), 122–134.

30 Haller, G. (2001) Distinguished material surfaces and coherent structures in three-dimensional fluid flows. *Physica D*, **149** (4), 248–277.

31 Haller, G. (2002) Lagrangian coherent structures from approximate velocity data. *Physics of Fluids*, **14** (6), 1851–1861.

32 Shadden, S.C., Lekien, F., and Marsden, J.E. (2005) Definition and properties of Lagrangian coherent structures from finite-time Lyapunov exponents in two-dimensional aperiodic flows. *Physica D*, **212** (3–4), 271–304.

33 Lekien, F., Shadden, S.C., and Marsden, J.E. (2007) Lagrangian coherent structures in n-dimensional systems. *Journal of Mathematical Physics*, **48** (6), 065404.

34 Haller, G. (2000) Finding finite-time invariant manifolds in two-dimensional velocity fields. *CHAOS*, **10** (1), 99–108.

35 Haller, G. (2001) Lagrangian structures and the rate of strain in a partition of two-dimensional turbulence. *Physics of Fluids A*, **13**, 3368–3385.

36 Aurell, E., Boffetta, G., Crisanti, A., Paladin, G., and Vulpiani, A. (1997) Predictability in the large: An extension of the concept of Lyapunov exponent. *Journal of Physics A – Mathematical and General*, **30** (1), 1–26.

37 Boffetta, G., Lacorata, G., Redaelli, G., and Vulpiani, A. (2001) Detecting barriers to transport: A review of different techniques. *Physica D*, **159**, 58–70.

38 Joseph, B. and Legras, B. (2002) Relation between kinematic boundaries, stirring, and barriers for the antarctic polar vortex. *Journal of the Atmospheric Sciences*, **59** (7), 1198–1212.

39 Koh, T.Y. and Legras, B. (2002) Hyperbolic lines and the stratospheric polar vortex. *CHAOS*, **12** (2), 382–394.

40 Lekien, F. (2003) Time-dependent dyanmical systems and geophysical flows Ph.D., California Institute of Technology.

41 Branicki, M. and Wiggins, S. (2010) Finite-time Lagrangian transport analysis: stable and unstable manifolds of hyperbolic trajectories and finite-time Lyapunov exponents. *Nonlinear Processes in Geophysics*, **17** (1), 1–36.

42 Shadden, S.C., Dabiri, J.O., and Marsden, J.E. (2006) Lagrangian analysis of fluid transport in empirical vortex ring flows. *Physics of Fluids*, **18** (4), 047105.

43 Shadden, S.C., Katija, K., Rosenfeld, M., Marsden, J.E., and Dabiri, J.O. (2007) Transport and stirring induced by vortex formation. *Journal of Fluid Mechanics*, **593**, 315–331.

44 Wiggins, S. (2005) The dynamical systems approach to Lagrangian transport in oceanic flows. *Annual Review of Fluid Mechanics*, **37**, 295–328.

45 Coppel, W.A. (1978) *Dichotomies in Stability Theory*, Springer, Berlin; New York.

46 Mathur, M., Haller, G., Peacock, T., Ruppert-Felsot, J.E., and Swinney, H.L. (2007) Uncovering the Lagrangian skeleton of turbulence. *Physical Review Letters*, **98** (14), 144502.

47 Beron-Vera, F.J., Olascoaga, M.J., Brown, M.G., Kocak, H., and Rypina, I.I. (2010) Invariant-tori-like Lagrangian coherent structures in geophysical flows. *CHAOS*, **20** (1), 017514.

48 Shadden, S.C., Astorino, M., and Gerbeau, J.-F. (2010) Computational analysis of an aortic valve jet with Lagrangian coherent structures. *CHAOS*, **20** (1), 017512.

49 Jimenez, R. (2011) CUDA LCS Software http://www.its.caltech.edu/raymondj/LCS/.

50 du Toit, P., Newman, http://www.cds.caltech.edu/pdutoit/Philip_Du_Toit/Software.html (Last access Nov. 12, 2010).

51 Brunton, S.L. and Rowley, C.W. (2010) Fast computation of finite-time Lyapunov exponent fields for unsteady flows. *CHAOS*, **20** (1), 017503.

52 Tabor, M. and Klapper, I. (1994) Stretching and alignment in chaotic and turbulent flows. *Chaos, Solitons & Fractals*, **4**, 1031–1055.

53 Shadden, S.C. (2006) A dynamical systems approach to unsteady systems Ph.D., California Institute of Technology.

54 Sadlo, F. and Peikert, R. (2007) efficient visualization of lagrangian coherent structures by filtered AMR ridge extraction. *IEEE Transactions on Visualization and Computer Graphics*, **13** (5), 1456–1463.

55 Sadlo, F., Rigazzi, A., and Peikert, R. (2010) Time-dependent visualization of lagrangian coherent structures by grid advection, in *Topological Methods in Data Analysis and Visualization* (eds. V. Pascucci, X. Tricoche, H. Hagen, and J. Tierny), Springer, Berlin, pp. 151–165.

56 Garth, C., Gerhardt, F., Tricoche, X., and Hagen, H. (2007) Efficient computation and visualization of coherent structures in fluid flow applications. *IEEE Transactions on Visualization and Computer Graphics*, **13** (6), 1464–1471.

57 Boffetta, G., Celani, A., Cencini, M., Lacorata, G., and Vulpiani, A. (2000) Nonasymptotic properties of transport and mixing. *CHAOS*, **10** (1), 50–60.

58 Lekien, F. and Leonard, N. (2004) *Dynamically consistent lagrangian coherent structures*, in 8th Experimental Chaos Conference, volume 742 of AIP Conference Proceedings (eds. S. Boccaletti, B.J. Gluckman, J. Kurths, L.M. Pecora, R. Meucci, and O. Yordanov), Florence, Italy, pp. 132–139.

59 Tang, W., Chan, P.W., and Haller, G. (2010) Accurate extraction of Lagrangian coherent structures over finite domains with application to flight data analysis over Hong Kong International Airport. *CHAOS*, **20** (1), 017502.

60. Haller, G. and Sapsis, T. (2011) Lagrangian coherent structures and the smallest finite-time Lyapunov exponent. *CHAOS* (preprint).

61. Li, G.S., Tricoche, X., and Hansen, C.D. (2006) GPUFLIC: Interactive and accurate dense visualization of unsteady flows, in *Eurographics/IEEE-VGTC Symposium on Visualization* (eds. T. Ertl, K. Joy, and B. Santo), The Eurographics Association, pp. 29–34.

62. Shadden, S.C., Lekien, F., Paduan, J.D., Chavez, F.P., and Marsden, J.E. (2009) The correlation between surface drifters and coherent structures based on high-frequency radar data in Monterey Bay. *Deep-Sea Research Part Ii-Topical Studies in Oceanography*, **56** (3–5), 161–172.

63. Lipinski, D. and Mohseni, K. (2010) A ridge tracking algorithm and error estimate for efficient computation of Lagrangian coherent structures. *CHAOS*, **20** (1), 017504.

64. Ferstl, F., Buerger, K., Theisel, H., and Westermann, R. (2010) Interactive separating streak surfaces. *IEEE Transactions on Visualization and Computer Graphics*, **16** (6), 1569–1577.

65. Olcay, A.B., Pottebaum, T.S., and Krueger, P.S. (2010) Sensitivity of Lagrangian coherent structure identification to flow field resolution and random errors. *CHAOS*, **20** (1)

66. Harrison, C.S. and Glatzmaier, G.A. (2011) Lagrangian coherent structures in the California current system - sensitivities and limitations. *Geophysical and Astrophysical Fluid Dynamics*, doi: 10.1080/03091929.2010.532793.

67. Duran-Matute, M. and Velasco Fuentes, O.U. (2008) Passage of a barotropic vortex through a gap. *Journal of Physical Oceanography*, **38** (12), 2817–2831.

68. Resplandy, L., Levy, M., d'Ovidio, F., and Merlivat, L. (2009) Impact of submesoscale variability in estimating the air-sea CO_2 exchange: Results from a model study of the POMME experiment. *Global Biogeochemical Cycles*, **23**, GB1017.

69. Rypina, I.I., Brown, M.G., and Beron-Vera, F.J. (2007) On the lagrangian dynamics of atmospheric zonal jets and the permeability of the stratospheric polar vortex. *Journal of the Atmospheric Sciences*, **64** (10), 3595–3610.

70. Lekien, F. and Ross, S.D. (2010) The computation of finite-time Lyapunov exponents on unstructured meshes and for non-Euclidean manifolds. *CHAOS*, **20** (1), 017505.

71. de la Camara, A., Mechoso, C.R., Ide, K., Walterscheid, R., and Schubert, G. (2010) Polar night vortex breakdown and large-scale stirring in the southern stratosphere. *Climate Dynamics*, **35** (6), 965–975.

72. Lekien, F. and Coulliette, C. (2007) Chaotic stirring in quasi-turbulent flows. *Philosophical Transactions of the Royal Society A – Mathematical Physical and Engineering Sciences*, **365** (1861), 3061–3084.

73. d'Ovidio, F., Isern-Fontanet, J., Lopez, C., Hernandez-Garcia, E., and Garcia-Ladon, E. (2009) Comparison between Eulerian diagnostics and finite-size Lyapunov exponents computed from altimetry in the Algerian basin. *Deep-Sea Research Part I-Oceanographic Research Papers*, **56** (1), 15–31.

74. Rypina, I.I., Brown, M.G., and Kocak, H. (2009) Transport in an idealized three-gyre system with application to the adriatic sea. *Journal of Physical Oceanography*, **39** (3), 675–690.

75. Gildor, H., Fredj, E., Steinbuck, J., and Monismith, S. (2009) Evidence for submesoscale barriers to horizontal mixing in the ocean from current measurements and aerial photographs. *Journal of Physical Oceanography*, **39** (8), 1975–1983.

76. Rossi, V., Lopez, C., Hernandez-Garcia, E., Sudre, J., Garcon, V., and Morel, Y. (2009) Surface mixing and biological activity in the four Eastern Boundary Upwelling Systems. *Nonlinear Processes in Geophysics*, **16** (4), 557–568.

77. Branicki, M. and Malek-Madani, R. (2010) Lagrangian structure of flows in the Chesapeake Bay: challenges and

perspectives on the analysis of estuarine flows. *Nonlinear Processes in Geophysics*, **17** (2), 149–168.

78 Carlson, D.F., Fredj, E., Gildor, H., and Rom-Kedar, V. (2010) Deducing an upper bound to the horizontal eddy diffusivity using a stochastic Lagrangian model. *Environmental Fluid Mechanics*, **10** (5), 499–520.

79 Rypina, I.I., Pratt, L.J., Pullen, J., Levin, J., and Gordon, A.L. (2010) Chaotic advection in an archipelago. *Journal of Physical Oceanography*, **40** (9), 1988–2006.

80 Beron-Vera, F.J., Olascoaga, M.J., and Goni, G.J. (2010) Surface ocean mixing inferred from different multisatellite altimetry measurements. *Journal of Physical Oceanography*, **40** (11), 2466–2480.

81 Beron-Vera, F.J. (2010) Mixing by low- and high-resolution surface geostrophic currents. *Journal of Geophysical Research-Oceans*, **115**.

82 Reniers, A.J.H.M., MacMahan, J.H., Beron-Vera, F.J., and Olascoaga, M.J. (2010) Rip-current pulses tied to Lagrangian coherent structures. *Geophysical Research Letters*, **37**, L05605.

83 Beron-Vera, F.J. and Olascoaga, M.J. (2009) An assessment of the importance of chaotic stirring and turbulent mixing on the West Florida Shelf. *Journal of Physical Oceanography*, **39** (7), 1743–1755.

84 Olascoaga, M.J., Beron-Vera, F.J., Brand, L.E., and Kocak, H. (2008) Tracing the early development of harmful algal blooms on the West Florida Shelf with the aid of Lagrangian coherent structures. *Journal of Geophysical Research-Oceans*, **113** (C12), C12014.

85 Beron-Vera, F.J., Olascoaga, M.J., and Goni, G.J. (2008) Oceanic mesoscale eddies as revealed by Lagrangian coherent structures. *Geophysical Research Letters*, **35** (12), L12603.

86 Olascoaga, M.J., Rypina, I.I., Brown, M.G., Beron-Vera, F.J., Kocak, H., Brand, L.E., Halliwell, G.R., and Shay, L.K. (2006) Persistent transport barrier on the West Florida Shelf. *Geophysical Research Letters*, **33** (22), L22603.

87 Tang, W. and Peacock, T. (2010) Lagrangian coherent structures and internal wave attractors. *CHAOS*, **20** (1), 017508.

88 Rutherford, B., Dangelmayr, G., Persing, J., Schubert, W.H., and Montgomery, M.T. (2010) Advective mixing in a nondivergent barotropic hurricane model. *Atmospheric Chemistry and Physics*, **10** (2), 475–497.

89 Rutherford, B., Dangelmayr, G., Persing, J., Kirby, M., and Montgomery, M.T. (2010) Lagrangian mixing in an axisymmetric hurricane model. *Atmospheric Chemistry and Physics*, **10** (14), 6777–6791.

90 Coulliette, C., Lekien, F., Paduan, J.D., Haller, G., and Marsden, J.E. (2007) Optimal pollution mitigation in monterey bay based on coastal radar data and nonlinear dynamics. *Environmental Science & Technology*, **41** (18), 6562–6572.

91 Tang, W., Haller, G., Baik, J.-J., and Ryu, Y.-H. (2009) Locating an atmospheric contamination source using slow manifolds. *Physics of Fluids*, **21** (4), 043302.

92 Shadden, S.C., Katija, K., Rosenfeld, M., Marsden, J.E., and Dabiri, J.O. (2007) Transport and stirring induced by vortex formation. *Journal of Fluid Mechanics*, **593**, 315–331.

93 Olcay, A.B. and Krueger, P.S. (2008) Measurement of ambient fluid entrainment during vortex ring formation. *Experiments in Fluids*, **44**, 235–247.

94 Olcay, A.B. and Krueger, P.S. (2008) Measurement of ambient fluid entrainment during laminar vortex ring formation. *Experiments in Fluids*, **44** (2), 235–247.

95 Green, M.A., Rowley, C.W., and Haller, G. (2007) Detection of Lagrangian coherent structures in three-dimensional turbulence. *Journal of Fluid Mechanics*, **572**, 111–120.

96 Lipinski, D., Cardwell, B., and Mohseni, K. (2008) A Lagrangian analysis of a two-dimensional airfoil with vortex shedding. *Journal of Physics A-*

Mathematical and Theoretical, **41** (34), 344011.
97 Eldredge, J.D. and Chong, K. (2010) Fluid transport and coherent structures of translating and flapping wings. *CHAOS*, **20** (1), 017509.
98 Yang, A., Jia, L., and Yin, X. (2010) Experimental study of a starting vortex ring generated by a thin circular disk. *Journal of Bionic Engineering*, **7** (Suppl S), S103–S108.
99 Cardwell, B.M. and Mohseni, K. (2008) Vortex shedding over a two-dimensional airfoil: Where the particles come from. *AIAA Journal*, **46** (3), 545–547.
100 Tang, W., Mathur, M., Haller, G., Hahn, D.C., and Ruggiero, F.H. (2010) Lagrangian coherent structures near a subtropical jet stream. *Journal of the Atmospheric Sciences*, **67** (7), 2307–2319.
101 Ruiz, T., Boree, J., Tran, T., Sicot, T.C., and Brizzi, L.E. (2010) Finite time Lagrangian analysis of an unsteady separation induced by a near wall wake. *Physics of Fluids*, **22** (7), 075103.
102 Lukens, S., Yang, X., and Fauci, L. (2010) Using Lagrangian coherent structures to analyze fluid mixing by cilia. *CHAOS*, **20** (1), 017511.
103 Peng, J., Dabiri, J.O., Madden, P.G., and Lauder, G.V. (2007) Non-invasive measurement of instantaneous forces during aquatic locomotion: A case study of the bluegill sunfish pectoral fin. *Journal of Experimental Biology*, **210** (4), 685–698.
104 Peng, J. and Dabiri, J.O. (2007) A potential-flow, deformable-body model for fluid–structure interactions with compact vorticity: Application to animal swimming measurements. *Experiments in Fluids*, **43** (5), 655–664.
105 Franco, E., Pekarek, D.N., Peng, J., and Dabiri, J.O. (2007) Geometry of unsteady fluid transport during fluid–structure interactions. *Journal of Fluid Mechanics*, **589**, 125–145.
106 Peng, J. and Dabiri, J.O. (2008) An overview of a Lagrangian method for analysis of animal wake dynamics. *Journal of Experimental Biology*, **211** (2), 280–287.
107 Peng, J. and Dabiri, J.O. (2008) The 'upstream wake' of swimming and flying animals and its correlation with propulsive efficiency. *Journal of Experimental Biology*, **211** (16), 2669–2677.
108 Lipinski, D. and Mohseni, K. (2009) Flow structures and fluid transport for the hydromedusae Sarsia tubulosa and Aequorea victoria. *Journal of Experimental Biology*, **212** (15), 2436–2447.
109 Wilson, M.M., Peng, J., Dabiri, J.O., and Eldredge, J.D. (2009) Lagrangian coherent structures in low Reynolds number swimming. *Journal of Physics-Condensed Matter*, **21** (20), 204105.
110 Green, M.A., Rowley, C.W., and Smits, A.J. (2010) Using hyperbolic Lagrangian coherent structures to investigate vortices in bioinspired fluid flows. *CHAOS*, **20** (1), 017510.
111 Shadden, S.C. and Taylor, C.A. (2008) Characterization of coherent structures in the cardiovascular system. *Annals of Biomedical Engineering*, **36** (7), 1152–1162.
112 Xu, Z., Chen, N., Shadden, S.C., Marsden, J.E., Kamocka, M.M., Rosen, E.D., and Alber, M. (2009) Study of blood flow impact on growth of thrombi using a multiscale model. *Soft Matter*, **5** (4), 769–779.
113 Vetel, J., Garon, A., and Pelletier, D. (2009) Lagrangian coherent structures in the human carotid artery bifurcation. *Experiments in Fluids*, **46** (6), 1067–1079.
114 Santitissadeekorn, N., Bohl, D., and Bollt, E.M. (2009) Analysis and modeling of an experimental device by finite-time lyapunov exponent method. *International Journal of Bifurcation and Chaos*, **19** (3), 993–1006.

4
Interfacial Transfer from Stirred Laminar Flows
Joseph D. Kirtland and Abraham D. Stroock

4.1
Introduction

Interfacial transport is essential to the success of many processes that operate under laminar flow conditions in both macro- and microscale systems. Heat and mass transfer to solid–liquid interfaces and across liquid–liquid interfaces are fundamental to heat exchanger design [1–4], electrochemical systems for analysis and energy production [5–7], separations with membranes [7] and without membranes [8], fabrication at fluid–fluid interfaces [9, 10], and sensors involving interfacial reactions [11–15]. Due to the small characteristic dimension of the conduits in microfluidic systems, for typical solvents, microfluidic flows are characterized by low Reynolds number, $Re = UH/\nu < 100$, where U is the average speed of the flow, $H = 10\text{--}10^3$ μm is the characteristic dimension of the system, and ν is the kinematic viscosity of the fluid. In this regime, flows are laminar, inertial effects are weak, and turbulence does not occur. Despite the small dimensions, Péclet numbers can be large, in particular, for mass transfer: $Pe_M = Sc Re > 100$ where the Schmidt number $Sc = \nu/D = 10^3\text{--}10^5$ and D is the diffusivity of the solute of interest (for heat, $Pe_H = Pr Re < 10^3$ where the Prandtl number, $Pr = \nu/\alpha < 10$, and α is the thermal diffusivity of the fluid). At high Pe in laminar flows, rates of interfacial transfer can be dominated by diffusion such that one may require undesirably large interfacial areas or long transfer times to achieve the necessary exchange of heat or mass.

In this chapter, we review experimental and theoretical work related to the control of rates of interfacial transfer in microfluidic devices. We begin by presenting the phenomena and definitions of important parameters. We proceed with a presentation of experiments that have been performed to characterize transfer rates in various microfluidic and macroscopic systems under laminar conditions. We then pass to a discussion of numerical and theoretical approaches that have been proposed to model these phenomena. As Péclet numbers tend to be higher for mass transfer than for heat transfer in liquid flows, we privilege the treatment of mass transfer processes; the discussion is nonetheless applicable to heat transfer as well. We use the scenario of single-phase liquid flow as our central example although we do present selected results for liquid–liquid, multiphase cases. As a final note on our emphasis, we attempt to clarify how different classes of flows (nonchaotic and chaotic) affect rates of interfacial transfer.

4.2
Phenomena and Definitions

In this section, we provide an overview of the phenomena observed in interfacial transfer, provide basic definitions, and pose questions that we will address later in the chapter. For this overview, we use the results of numerical calculations (Figure 4.1) of transfer in a specific yet common microfluidic scenario. In Sections 4.3 and 4.4, we will review other scenarios and provide further information and analysis on the predictions presented in Figure 4.1.

Figure 4.1a(i) presents a simple microfluidic scenario that involves interfacial mass transfer: a reaction occurs at one wall of a microchannel of rectangular

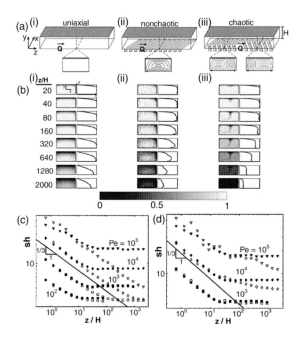

Figure 4.1 Interfacial transfer from microfluidic flows calculated via a Lagrangian method. (a) Schematic diagram of microchannels with reaction occurring at one boundary. A fast reaction occurs at the top boundary of the channel (gray) and a mean flow travels along z. (b) Distributions of reactive solute in the cross-section of a rectangular channel with uniaxial flow (i) and three-dimensional flows that exhibit nonchaotic advection (ii) and chaotic advection (iii). (c–d) Local Sherwood number for chaotic flow (filled symbols) compared to uniaxial flow (open symbols in c) and nonchaotic, 3D flow (open symbols in d). Adapted from [16].

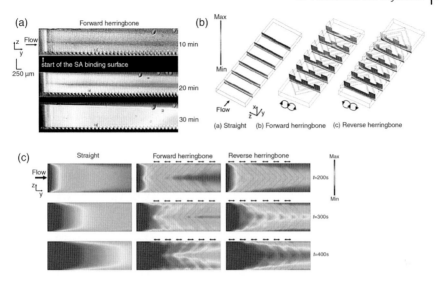

Figure 4.2 Protein binding. (a) Experimental surface concentration profile of bound streptavidin in the reverse herringbone microchannel and straight microchannel at time = 20 min. Images created by subtracting the image at 20 min from the initial image with only PBS present. Image is contrast-enhanced. Model concentration profile of streptavidin in each microchannel (mol m^{-3}). The maximum and minimum concentrations were 40 and 0 nM, respectively. The concentration profile in each microchannel is consistent with the spatial distribution of bound SA. Model surface concentration profile of bound streptavidin (mol m^{-2}) in each microchannel over time. The arrows indicate the deepest portion of the channel where the herringbone microstructure was present. The maximum and minimum concentrations were 3.99×10^{-8} and 0 mol m^{-2}, respectively. Note the very dramatic differences in the spatial patterns of surface binding. Adapted from [14].

cross-section. This reactive wall could be, for example, an electrode in an electrochemical system [5, 17] (see Figure 4.3) or the surface of a sensor for a binding event [14] (see Figure 4.2). In a channel with smooth, straight walls (Figure 4.1a(i)), a pressure-driven flow will be uniaxial. Figure 4.1b(i) shows numerical solutions for the evolution of the distribution of concentration of a reactive solute within the cross-section of such a channel for the case in which the reaction at the wall is instantaneous (concentration at reactive wall is zero). As one proceeds along the channel, a zone depleted of the reactive solute (the concentration boundary layer) grows into the channel from the reactive boundary; this boundary layer is thicker for a solute with higher diffusivity than for a solute with lower diffusivity. Solute must diffuse across this layer to reach the boundary, such that the diffusion-limited flux averaged across the width of the reactive boundary (along the x-axis), $J(z)$ (mole m^{-2} s^{-1}) is

$$J(z) = k(z)C(z) = \frac{D}{\delta(z)} C(z) \qquad (4.1)$$

where $C(z)$ (mole m^{-3}) is the velocity-weighted average (cup-mixing average) concentration in the cross-section at z, $k(z)$ (m s^{-1}) is the local mass transfer coefficient (averaged over the width, along x), and $\delta(z)$ (m) is the effective thickness of the boundary layer. The second equality in Eq. (4.1) defines $\delta(z)$. We define a nondimensional mass transfer coefficient (local along the axial dimension, z, and averaged over the width, x), the Sherwood number:

$$Sh(z) = \frac{Hk(z)}{D} = \frac{H}{\delta(z)} \qquad (4.2)$$

The equivalent quantity for heat transfer is the Nusselt number [19]. The open symbols in Figure 4.1c show the evolution of $Sh(z)$ at several values of Pe in this uniaxial case (open triangles). The rate of mass transfer decays as the boundary layer grows to fill the depth of the channel ($\delta \to H$) and then reaches an asymptotic value of order one, $Sh(z \to \infty) = O(1)$. The Sherwood number represents the ratio of the rate of transfer with a given flow to the rate in the absence of flow (pure diffusion within the same geometry). The challenge of maximizing rates of interfacial transfer can be expressed as that of maximizing $Sh(z)$. We also note that Eq. (4.2) provides a geometrical interpretation of this challenge, namely that one must minimize δ, the thickness of the boundary layer. The observed evolution – an entrance region with decaying Sh followed by an asymptotic region with $Sh(z)$ of order one – in this duct flow with a single reactive boundary is entirely analogous to the classic result of Grætz (1880s) for heat transfer from a Poiseuille flow to the walls of a circular pipe [19, 20].

The introduction of secondary flows that stir the fluid offers a means of decreasing the thickness of the boundary layer relative to the default, uniaxial case in microfluidic flows. Figure 4.1a(ii) and a(iii) shows examples of channel that present grooves on one wall. The grooves in the bottom of the channel generate secondary flows in the cross-section [21]. When the pattern of the grooves is constant along the channel (Figure 4.1a(ii)), the resulting flow is three-dimensional (3D) and nonchaotic; when the pattern varies, the flow is 3D and can be chaotic. Both cases can be modeled with the flow of a lid-driven cavity in which regions of parallel grooves are replaced with a region of flat wall with lateral slip (illustrated in the cross-section with streamlines – Figure 4.1a(ii) and a(iii)) [16, 22, 23]. As seen in numerical solutions for the distribution of concentration in these types of flows (Figure 4.1b(ii) and b(iii)), stirring the fluid in the cross-section deforms the depleted zone and reduces its thickness. Interestingly, as in the uniaxial case, the distribution appears to evolve toward an asymptotic form. This evolution can be seen quantitatively in $Sh(z)$: for the chaotic case (filled symbols in Figure 4.1c and d) and the nonchaotic case (open symbols in Figure 4.1d), $Sh(z)$ drops initially, closely following the decay of the uniaxial case (open symbols in Figure 4.1c). Beyond this entrance region, the behavior in the chaotic and nonchaotic flows (Figure 4.1d) differs in a subtle way: in the chaotic case, $Sh(z)$ reaches a constant value whereas in the nonchaotic it continues to decay slowly; both values remain of the same order of magnitude. We note the subtlety of this difference between the interfacial transfer in chaotic and the nonchaotic flows is in sharp contrast with the dramatic difference that is observed for mixing between these two classes of flows [21, 24].

4.3
Experimental Methods

From this introduction to the phenomenon, we can pose more precise questions regarding the manipulation of interfacial transfer. In particular, we aim to understand how the character of the flow – uniaxial, multidimensional and nonchaotic, or multidimensional and chaotic – influence the rates of interfacial transfer, and to express this understanding as a prediction for the form of $Sh(z)$ as a function of simple parameters that characterize the flow.

4.3
Experimental Methods

Figures 4.2–4.5 present various experimental systems in which measurements of interfacial mass transfer (Figures 4.2, 4.3, and 4.5) and heat transfer (Figure 4.4) have been studied in laminar flows. These examples provide a vision of the technological contexts in which interfacial transfer is important. They also provide a sense of the experimental strategies that can be employed to extract quantitative measurements of rates of transfer. When appropriate, we have included modeling studies of these systems. We discuss these models in the next section.

4.3.1
Protein Binding

Figure 4.2a presents the spatially resolved measurement of the quantity of bound protein on a surface on the floor of a microchannel that presented a symmetric herringbone pattern of grooves [14]. The measurement was performed optically via surface plasmon resonance (SPR). The geometry of the microchannel is shown schematically in the middle diagram of Figure 4.2b. This experiment is unusual in its ability to provide both temporally and spatially resolved information about the flux to the reactive boundary in the form of the rate of change of the local signal from bound protein. We note the emergence of spatially inhomogeneous patterns at the boundary. The dark zone in the center of the channel corresponds to the upwelling in the secondary recirculation generated by the grooves. This zone corresponds to the thick region of the boundary layer that is visible in Figure 4.1b. The qualitative characteristics of the pattern are captured by the numerical calculations presented in the center frame in Figure 4.2c. The trade-off with this spatial resolution is that it is challenging to follow the process over large sections of channel and thus to study both the entrance region and the asymptotic zone, especially at higher values of Pe.

4.3.2
Electrochemical Reactions

Figure 4.3a presents a microfluidic electrochemical system with which we have studied mass transfer from various laminar flows. This potential cell [26] provides access to the global flux to the boundaries via the current collected at electrodes on the boundaries of the channels. Figure 4.3b shows a scanning electron micrograph of an

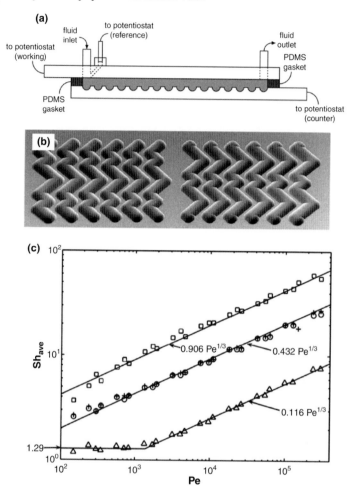

Figure 4.3 Electrochemical measurements of rates of mass transfer in microfluidic potential cells. (a) Schematic diagram of microfluidic potential cell. Top and bottom walls of channel act as working and counter electrodes, respectively. Grooves of herringbone mixing structure are indicated on bottom wall. (b) Scanning electron micrograph of counter electrode with groves in staggered herringbone motif. (c) Globally averaged Sherwood numbers evaluated via Eq. (4.3) from measured current for uniaxial flow over smooth counter electrode (triangles), 3D, chaotic flow over counter with staggered herringbone motif (crosses), and 3D, nonchaotic flow over counter with symmetrical herringbone motif (circles). Experiments were run with the reversible redox species, ferricyanide/ferrocyanide. Adapted from [18].

electrode fabricated with the staggered herringbone pattern of grooves in glass. With the use of a fast, reversible redox couple, one can adjust the voltage to operate in the mass-transfer limited regime. From the measured current, I (A) one can extract the globally averaged Sherwood number, which, by integration of Eq. (4.2), has the form [16]

$$Sh_{ave} = \int_0^L Sh(z)dz = \frac{PeH}{L}\frac{I}{mFC_0Q} \quad (4.3)$$

where L is the length of the electrodes, m is the number of electrons transferred per redox reaction, F (C mole^{-1}) is Faraday's constant, C_0(mole m^{-3}) is the concentration at the inlet, and Q (m^3 s^{-1}) is the volumetric flow rate. Figure 4.3c presents $Sh_{ave}(Pe)$ for three cases: (i) smooth-walled channel, (ii) grooved channel with parallel pattern that generates nonchaotic flow (as in Figure 4.1a, middle), and (iii) grooved channel with staggered herringbone pattern that generates chaotic flow (as in Figure 4.1a, right). In the cases with grooves, the working electrode to which the transport-limited flux occurred was opposite the grooved-boundary, as illustrated in Figure 4.1a. We remark that in all cases, Sh_{ave} follows robust power laws at high Pe: $Sh_{ave} \sim Pe^{1/3}$. We also note that the presence of the secondary flows provides a significant increase in the rates of mass transfer (>3-fold) and this increase is nearly identical for the chaotic and nonchaotic cases. We will return to a discussion of these observations in the section on modeling. A key advantage of electrochemical methods for the study of interfacial mass transfer stems from the ability to make quantitative measurements of rates. A disadvantage of the experiment presented in Figure 4.3 is that the use of continuous electrodes allowed for only global measurements of flux. Segmented electrodes could offer spatially resolved measurements.

4.3.3
Heat Transfer from Macroscopic Coiled Pipe

Figure 4.4 presents a macroscopic experiment in which heat transfer is measured in laminar flows at intermediate Reynolds numbers [3]. These authors and others [27] have demonstrated that the alternating axis coiled pipe pictured in Figure 4.4a generates chaotic trajectories due to the reorientation of secondary Dean flows (Figure 4.4b) in the bends as a function of axial position; laminar flow in a helical pipe is nonchaotic. We note that these Dean flows, despite their dependence on inertial effects at finite Reynolds number ($Re > 10$), have been exploited to stir flows in microfluidic devices [28]. In structures such as these, the authors have measured global rates of heat transfer by measuring the temperature change from the entrance to the exit of the pipe in a tube-in-shell configuration. Figure 4.4c presents the efficiency of heat transfer based on this global temperature change. Although further analysis was not performed in this study, we note that the efficiency reported, ε, can be related to the global Sherwood number (typically referred to as the Nusselt number in heat transfer) as follows:

$$Sh_{ave} = -\frac{PeR}{L}\ln(1-\varepsilon) \quad (4.4)$$

where R and L are the radius and length of the pipe, respectively. We note that the efficiency of heat transfer in this system is substantially larger (20–30%) for the chaotic flow relative to the nonchaotic one ("helical" in Figure 4.4c). Nonetheless, the two rates are of the same order of magnitude.

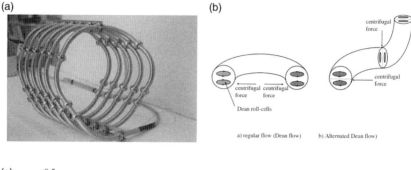
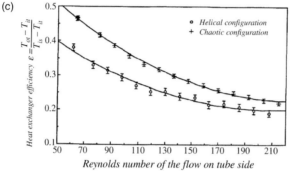

Figure 4.4 Heat transfer from macroscopic coiled pipes. (a) Chaotic heat exchanger (alternating axis coiled pipe). (b) Representation of streamlines associated with Dean's flows inside helical coiled pipe (left) and alternating axis coiled pipe (right). (c) Global efficiency of heat transfer as a function of Reynolds number for a helical pipe (nonchaotic) and an alternating axis coiled pipe (chaotic). Adapted from (Mokrani).

4.3.4
Interphase Mass Transfer from Droplets

Figure 4.5 presents interfacial transfer in a liquid–liquid, multiphase flow in a microfluidic system [25]. The channel geometries presented in Figure 4.5a allow for the formation of droplets of an aqueous phase within an organic continuous phase. In their experiments, Mary et al. followed the transfer of fluorescent solutes between these phases (Figure 4.5b – fluorescence only shown within droplets). They refer to the process as extraction when the solute passes from the continuous phase into the droplets and purification when it passes from the droplet into the continuous phase. The secondary flows are nonchaotic for straight channels and can be chaotic for serpentine channels [29]. As is seen in Figure 4.5b, this recirculation leads to the development of boundary layers within the droplets. For the steady nonchaotic flows shown, this boundary layer eventually encircles the entire droplet and reconnects with itself (for example, in the fourth droplet down channel). Figure 4.5c presents the evolution of the thickness of this boundary layer as a function of the residence time, $t = z/U$ of the droplets within the continuous phase. We note that the thickness

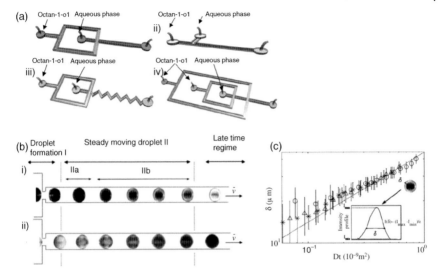

Figure 4.5 Interphase mass transfer from droplets in microfluidic flows. (a) Schematic views of the microfluidic chips for liquid–liquid extraction. (b) Fluorescence images of droplets at different times in the channel, respectively, (i) extraction, droplet composed of a mixture of water/60 wt% glycerol, and (ii) purification, water droplet. Note that distribution of fluorescence was only recorded in the region occupied by the droplets. (c) Experimentally measured thickness of the boundary layer as a function of Dt, where D is the molecular diffusivity of the solute and t is the residence time of the droplet in the flow. Adapted from [25].

remains relatively constant and then grows like, $\delta \sim (Dt)^{1/2}$. This behavior resembles that seen in Figure 4.1b for the nonchaotic case (center). Exploiting the serpentine channel (Figure 4.5a(iii)), the authors observed transfer in the presence of chaotic flow within the droplets; they note that the transfer rates were similar to those measured in nonchaotic flow. We note in passing that liquid–liquid exchange between laminar streams of miscible solvents has also been studied extensively for uniaxial flows [8].

4.3.5
Summary of Experimental Observations

Based on this brief review of experimental measurements of interfacial transfer in laminar flows, we note a few important observations: (1) the emergence of a boundary layer in the distribution of the diffusing specifies appears to be a general phenomenon (Figures 4.2b, (4.5)b and c). (2) The variation of key parameters appears to exhibit power law scaling with the residence time and flow rate; for transfer across no-slip interfaces the exponent is 1/3 (Figure 4.3c) and for transfer across slipping or stress-free interfaces it is 1/2 (Figure 4.5c). (3) Unlike for mixing processes, the difference in rates of interfacial transfer in chaotic and nonchaotic flows is small and the rates show qualitatively similar trends (for example, in scaling with flow parameters). We will attempt to explain these observations in the following section.

4.4
Modeling Approaches

We now turn to a review of approaches to model interfacial transfer processes in laminar flows. The central goal of these approaches is to provide predictions of rates of transfer (Sh) with a minimum of information about the flow. More precisely, we would like to find expressions for Sh(z) based on simple parameters of the Eulerian flow, such as the average flow speed or shear rate, the mass or thermal diffusivity, and the geometry. Given the increased computational capacities of today's computers, one can also attempt to obtain complete solutions for the distribution of concentration (or temperature) from which rates of transfer can be calculated. This detailed information can be important when, for example, deviations from the spatially averaged behavior (as in Figure 4.2c) may have important consequences on the system. In this section, we begin with a presentation of these numerical approaches with a summary of their predictions and a discussion of their advantages and limitations. We then proceed to present approximate analytical results that approach the goal of providing predictions of interfacial transfer rates with the least possible computational effort and provide mechanistic insights into how different types of flows affect transfer.

4.4.1
Numerical Solutions in Eulerian Frame

The governing equation for mass transfer in a flow field, $\vec{U}(x, y, z)$, is the convection diffusion equation

$$\frac{\partial C}{\partial t} + \vec{U} \cdot \vec{\nabla} C = D \nabla^2 C \tag{4.5}$$

The information about interfacial mass transfer is incorporated into Eq. (4.5) via boundary conditions on the concentration. For the example of an instantaneous reaction in Figure 4.1, the concentration at the reactive boundary is zero and the gradient normal to the nonreactive boundaries is zero. Equation (4.5) can be solved numerically for $C(x,y,z)$ in a specific geometry by discrete approaches such as the finite element and finite difference methods [14, 25, 30]. The local flux can then be extracted at the boundary of interest from Fick's first law

$$J_{rxn} = D\vec{n} \cdot \vec{\nabla} C \tag{4.6}$$

where $\vec{n}$ is the local normal to the boundary. In Equation 4.6, J_{rxn} is positive for flux into the fluid from the surface.

Figure 4.2b and c presents the predictions of a calculation by the finite element method (FEM) for a protein binding reaction at the grooved boundary of a microchannel from a steady flow: $Re = 0.0111$ and $Pe = 148$. In this calculation, the Navier–Stokes equations were solved for the flow field in tandem with Eq. (4.5) for the concentration field. In Figure 4.2b, we observe the development of the depletion boundary layer, as in Figure 4.1b. In Figure 4.2c, we observe the time evolution of the bound quantity of the molecule, calculated as the time integral of the local value of the flux in Eq. (4.6). We note qualitative agreement between the calculated flux for the "forward herringbone" and the spatial distribution of bound protein observed in the experiments at early times (Figure 4.2a). The authors observed that, in this entrance region of the channel, the rate of transfer was similar with and without the secondary flows generated by the grooves. This observation is consistent with the evolution of $Sh(z)$ in Figure 4.1c and d in which the values of $Sh(z)$ are the same for stirred and unstirred cases at the beginning of the channels; we will look for an explanation of this phenomenon in Section 4.4.4.

An advantage of a complete numerical solution stems from the ability to account explicitly for the geometrical details of the structure and their impact on the flow and rates of transfer. A disadvantage is that, with commonly available numerical packages and computers, the spatial and temporal extents of the calculations are subject to limitations due to limits on memory and computational power; in the study by Foley et al., only the entrance region of the process was explored. It may be possible to circumvent these limitations for periodic or unmodulated flow fields by imposing periodic boundary conditions in an iterative manner: that is, applying the concentration field at the outlet as a new inlet condition and recalculating, iteratively progressing down the channel. Such a methodology, although computationally intensive, would allow the study of arbitrary lengths of device and should permit the prediction of the asymptotic state as the inlet and outlet become self-similar.

4.4.2
Numerical Solutions in Lagrangian Frame

An alternative method of solving Eq. (4.5) involves calculating the trajectories of individual particles in the flow [16, 31, 32]. The trajectories of diffusive tracers follow

$$\frac{d\vec{x}}{dt} = \vec{U} + \vec{B}(t) \tag{4.7}$$

where $\vec{x}$ is the instantaneous position of the particle and $\vec{B}$ is a Gaussian distributed random displacement that represents Brownian motion. In a channel geometry, to simulate a steady inflow of uniform concentration, one starts trajectories from $z=0$ with a random distribution of xy-positions weighted by the velocity profile (for example, a Poiseuille profile for pressure-driven flow). Forward integration of Eq. (4.7) via, for example, a Rugga–Kutta scheme, gives the convective–diffusive trajectory of the particle down the channel. To account for interfacial transfer, one stops the trajectory when it encounters a reactive boundary and records the position of the encounter to construct a map of the flux; at nonreactive boundaries, one performs a specular reflection. The local Sherwood number can be evaluated from the flux or as

$$Sh(z) = -PeH\frac{d}{dz}\ln(C(z)) \tag{4.8}$$

Figure 4.1 presents the results of calculations performed by this method with an approximate flow field that mimics that generated by the grooves [16]. We point out again here that these calculations predicted (Figure 4.1c and d) that in all three classes of flows – uniaxial, 3D nonchaotic, and 3D chaotic – $Sh(z)$ undergoes the same initial decay. Furthermore, this decay has the form, $Sh(z) \sim (z/(PeH))^{-1/3}$; this prediction matches the behavior in the classic Grætz result [19]. For the two stirred flows, the evolution of $Sh(z)$ transitions away from this decay at a Pe-independent distance down the channel, whereas for the uniaxial case (and the Grætz result) the entrance length, $z_{ent} \sim PeH$. For the chaotically stirred flow, the transition is to a constant, Pe-dependent value of $Sh(z)$, $Sh_{asy} \sim Pe^{1/3}$ [16]. For the nonchaotic case, $Sh(z)$ continues to decay, but more slowly.

Figure 4.6 presents a study by Ghosh et al. in which they used Brownian tracers to calculate transfer into a separation zone (Figure 4.6a) in an unbounded flow [31]. Oscillations of the position of the separation zone leads to chaotic trajectories in this flow; the parameter β controls the amplitude of this oscillation. Figure 4.6b presents

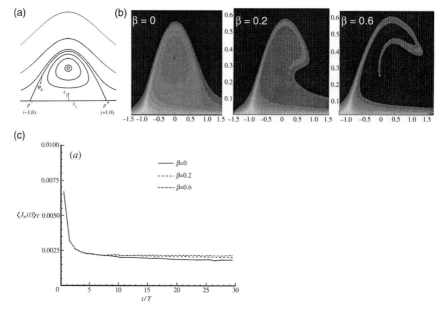

Figure 4.6 Interfacial transfer from a wall into a separation bubble. (a) Streamlines of a steady separation bubble formed in a uniform, unbounded flow by a no-slip region on a wall. Oscillations of this bubble lead to chaotic flow within the bubble. (b) Distribution of a diffusive scalar released from wall into a bubble in a chaotic regime. This distribution is a time-periodic asymptotic state. The parameter β is a measure of the amplitude of the oscillation. $\beta = 0$ corresponds to the steady, nonchaotic case. For $\beta = 0.2$ and 0.6, the flow contains chaotic islands that attach to boundary; these islands are larger for $\beta = 0.6$ and for $\beta = 0.2$. (c) Temporal evolution of the flux averaged over the extent of the bubble for a steady bubble (solid line – $\beta = 0$, nonchaotic) and two nonsteady bubbles (dashed lines – $\beta = 0.2$ and 0.6, chaotic). Adapted from [31].

the distribution of the scalar at late times. We note, in the case of the nonchaotic ($\beta = 0$) and weakly chaotic ($\beta = 0.2$) flows, the plume of scalar that was swept off of the wall has reconnected with the right edge of the separated zone; for the more strongly chaotic case ($\beta = 0.6$), the plume does not reconnect. Figure 4.6c presents the average flux to the boundary. We note first that the magnitude of the flux is similar for the chaotic and nonchaotic flows. Yet, only for the more chaotic case ($\beta = 0.6$) does the flux appear to reach a constant value; at late times for the nonchaotic ($\beta = 0$) and weakly chaotic ($\beta = 0.2$) cases, the flux continues to decay slowly.

Due to its inherently parallel nature (each trajectory can be calculated independently and requires little memory compared to the global solution in a finite element calculation), this method allows for the numerical study of transfer processes over large spatial and temporal domains. Importantly for the studies discussed here, this capacity gives access to the evolution of the concentration field all the way out to its asymptotic state, as illustrated in Figures 4.1 and 4.6. Further, we note that the integration of Eq. (4.7) with $\vec{B} = 0$ provides useful Lagrangian information about the flow, such as Poincaré maps and Lyapunov exponents [31–33]. Disadvantages of this method are that it requires a known flow field and is difficult to implement in complex geometries such as that presented by the grooves in the staggered herringbone mixer.

4.4.3
Macrotransport Approach

We mention briefly that Brenner et al. have implemented their macrotransport or generalized Taylor dispersion approach to study interfacial transfer in a variety of flows that exhibit both nonchaotic and chaotic behavior [34, 35]. In this approach, the convective–diffusive process within the flow is captured in a diffusion-like dispersion coefficient. Once this coefficient is evaluated, they can calculate the transfer of a scalar field from an averaged transport equation that does not require knowledge of the detailed structure of the flow or the three-dimensional concentration field. Employing this approach, they found that passage from a nonchaotic to a chaotic regime could increase the rate of transfer to a boundary by approximately 50%. The method is versatile but challenging to implement. This approach intrinsically considers only the asymptotic behavior of the system.

4.4.4
Theoretical Approaches

The Grætz analysis of interfacial transfer from a steady, uniaxial flow provides an exact solution of Eq. (4.5) (neglecting axial diffusion) from which the distribution of the scalar of interest and the flux can be evaluated everywhere [19]. This analysis captures the entrance region with appropriate scaling ($Sh(z) \sim (z/(PeH))^{-1/3}$) and the asymptotic region, $Sh(z \to \infty) = O(1)$ (Figure 4.1c). For more complicated flow fields (for example, with transverse secondary flows in the cross-section of the channel – Figure 4.1a(ii) and a(iii)), this exact analysis is intractable. In considering such flows, Chang and Sen proposed the use of boundary layers to reduce the problem to one of transfer through one or more resistors; the magnitude of these resistances depends on the effective thickness of the boundary layers through which the scalar must travel [36]. They identified the need to account for resistances both at boundaries of the flow and at streamlines that divide convectively separate regions of the flow. We adopted this perspective in Eqs. (4.1) and (4.2) with the introduction of the concept of a boundary layer (Section 4.2). This perspective is motivated qualitatively by the form of the distributions of concentration predicted by numerical approaches, as in Figures 4.1–4.3: gradients develop adjacent to the interface to which transfer occurs. This familiar phenomenon appears to be relevant even when flows in the bulk are chaotic. Indeed, we will argue below that these boundary layers are most robust in such flows. We note that "renewal theory" in the context of interphase mass transfer [37] provides precedent for this approach to the treatment of interfacial transfer in complex flows.

In order to make predictions with this approximate approach, one must analyze the development of the boundary layers. In considering this development, we identify two important cases related to the variation of the tangential velocity of the fluid at the interface across which transfer occurs: (1) constant tangential velocity or plug flow and (2) linearly varying tangential velocity or simple shear. The case of plug flow is relevant to transfer to stress-free interfaces (e.g., fluid–fluid or fluid–solid with perfect slip), moving no-slip boundaries, or internal streamlines along a separatrix in the flow. The case of shear flow is relevant to transfer to no-slip boundaries. Figure 4.7a and b presents these cases schematically. In these scenarios, analytical solutions can be found for the steady, two-dimensional, convection–diffusion equation

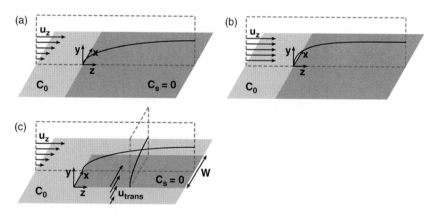

Figure 4.7 Development of boundary layers. (a) Growth of boundary layer into a fully developed, uniaxial flow over a no-slip boundary. (b) Growth of a fully boundary layer into a uniaxial flow over a stress-free boundary. (c) Conceptual picture of growth of a boundary layer into a multiaxial flow over a no-slip boundary with a principal component, u_z, and a secondary component, u_{trans} running in an orthogonal direction. The secondary flow runs along a bounded dimension of the transfer interface.

$$U_z \frac{\partial C}{\partial z} = D \frac{\partial^2 C}{\partial y^2} \tag{4.9}$$

Equation (4.9) is an approximate form of Eq. (4.5) in which we have neglected diffusion along the streamwise direction (a good approximation at high Pe) and adopted local axes with an origin at the leading edge of the interface at which transfer occurs. For the case of plug flow ($U_z = U =$ constant), we have via Eqs. (4.8) and (4.1) applied to analytical solutions of Eq. (4.7) [32],

$$k(z) = \frac{1}{2\sqrt{\pi}} \left(\frac{UD}{z}\right)^{1/2} \quad Sh(z) = \frac{1}{2\sqrt{\pi}} \left(\frac{z}{PeH}\right)^{-1/2} \tag{4.10}$$

Comparing Eqs. (4.10) and (4.2), we can identify an effective thickness of the depletion boundary layer:

$$\delta(z) = 2\sqrt{\pi} \left(\frac{zD}{U}\right)^{1/2} \tag{4.11}$$

The case of shear flow was treated by Lévêque [16, 38]:

$$k(z) = \frac{1}{\Gamma\left(\frac{4}{3}\right)} \left(\frac{\dot{\gamma} D^2}{9z}\right)^{1/3} \quad Sh(z) = \frac{1}{\Gamma\left(\frac{4}{3}\right)} \left(\frac{\dot{\gamma} H^3}{9Dz}\right)^{1/3} = \frac{1}{\Gamma\left(\frac{4}{3}\right)} \left(\frac{PeH}{9z}\right)^{1/3} \tag{4.12}$$

where $\dot{\gamma}(1\,\mathrm{s}^{-1})$ is the shear rate at the reactive interface and we have taken $Pe = \dot{\gamma} H^2/D$. We can again identify an effective boundary layer

$$\delta(z) = \Gamma\left(\frac{4}{3}\right)(9zD\dot{\gamma})^{1/3} \tag{4.13}$$

These are the standard results for uniaxial flows [19]. For example, Eq. (4.10) correctly predicts the evolution of $Sh(z)$ in the entrance region of the uniaxial channel flow, as indicated in Figure 4.1c. This model of a shear flow successfully treats the entrance behavior of interfacial transfer in a Poiseuille flow because in this region the boundary layer is still thin and thus only influenced by the simple shear adjacent to the boundary. Once this criterion is violated, the distribution of the scalar has nearly reached its asymptotic shape and $Sh(z)$ plateaus at a value of order one.

How though do these predictions relate to interfacial transfer in multidimensional flows, either chaotic or nonchoatic? A simple conceptual picture of the process is illustrated in Figure 4.7c for the case of a no-slip interface: a principal flow (along z) impinges on the leading edge of the unbounded axis of the interface and a boundary layer begins to grow, following Eq. (4.13). Simultaneously, a transverse, secondary flow impinges on the edge of the bounded x-axis of the interface; its development again follows Eq. (4.13) (with x in place of z). Importantly, for the simple reason that the transverse dimension of the interface is finite, the secondary boundary layer does not grow indefinitely. Rather, its growth is terminated once it has traversed the width, W. Once this transverse boundary layer has formed, it terminates the growth of the principal boundary layer as well, by sweeping the material off of the interface.

This process is visible in the cross-sections of the distribution of concentration in Figure 4.1b(ii) and b(iii); the depleted fluid is swept into the bulk of the flow. If the fluid that leaves the interface can be homogenized with the bulk (i.e., to the local cup mixing concentration) before returning, the system reaches an asymptotic state with a transfer rate defined by this initial transverse boundary layer. This case is illustrated by the chaotic case in Figure 4.1b(iii) and c. If, on the other hand, this homogenization does not take place, the boundary layer reconnects with the interface and continues to grow into the bulk; the rate of transfer rate continues to decay such that $Sh \rightarrow O(1)$. This behavior is seen for the nonchaotic cases in Figure 4.1b(ii) and d. For 2D flows (droplets in Figure 4.5 and separation zone in Figure 4.6), the important aspect of this argument is the existence of a transfer interface of finite length on which there must exist a stationary point at which the boundary layer detaches. The importance of the chaotic nature of the flow for the establishment of an asymptotic state is seen in Figure 4.6b and c: as noted in Section 4.4.2, with a robust chaotic flow ($\beta = 0.6$), the boundary layer does not reconnect and the rate of transfer remains constant and of (slightly) higher magnitude.

This simple picture of competing boundary layers with detachment also provides quantitative predictions. For the flows presented in Figures 4.1 and 4.3, this picture provides the following predictions [16, 32]: (1) in the presence of transverse, secondary flows, $Sh(z)$ decays in the same manner as the uniaxial flow (Eqs. (4.10) or (4.12)) out to a length that depends only on the ratio of the principal to the secondary flow speeds ($z_{ent} \sim UW/U_{trans}$), not on Pe. During this period, the transfer is controlled by the axial boundary layer. (2) Beyond this length in chaotic flows, $Sh(z)$ goes to a constant value that depends on the strength of the secondary flow:

$$Sh_{asy} = \frac{(3/4)^{1/3}}{\Gamma\left(\frac{4}{3}\right)} \left(\frac{H}{W}\right)^{1/3} \left(\frac{H^2 \dot{\gamma}_{trans}}{D}\right)^{1/3} \tag{4.14}$$

where $\dot{\gamma}_{trans}$ is the shear rate associated with the secondary flow at the interface. Together, these two predictions represent a complete correlation for $Sh(z)$. Due to the features that this behavior shares with the classic Grætz prediction (scaling in entrance region that transitions to a constant value beyond), we have termed it "modified Grætz" behavior. We note though that the particular characteristics of interfacial transfer in chaotic flows are favorable: the short and Pe-independent entrance length means that the efficiency of transfer varies only over a short zone within the flow; the Pe-dependent value of Sh_{asy} allows for this transfer to be made much larger than that in a uniaxial flow in a channel of similar dimensions.

Remaining questions relate to the differences between chaotic and nonchaotic flow. First, why do nonchaotic flows seem to fail to exhibit this modified Grætz behavior and is this observation general? By considering the fate of the fluid that leaves the transfer interface and enters the bulk, we have shown that it is the unusual character of mixing in chaotic flows that allows for homogenization of the boundary layer to occur before it can reconnect with the interface [32]. In typical nonchaotic flows, this homogenization occurs too slowly. Thus, the dramatic differences between mixing in chaotic and nonchaotic flows lead to subtle differences in rate of interfacial transfer. Second, why do chaotic and nonchaotic flows appear to provide very similar rates of transfer with similar scaling? This similarity arises due to the fact that typical measurements of transfer are dominated by the fluxes achieved at short distances (or early times). In this regime, the slow decay of the efficiency in nonchaotic flows at long times has little impact on the observed rate. We note that scenarios do exist in which the observed effectiveness of a process depends strongly on Sh_{asy}, as when the scalar that must be transferred is continuously produced in the fluid (e.g., exothermic reactions). In such cases, chaotic flow could provide much higher rates of transfer with qualitatively different dependence on Pe.

4.5
Conclusions

In conclusion, we stress that rates of interfacial transfer in laminar systems can be significantly increased by the introduction of multidimensional flows. Furthermore, the presence of chaotic advection leads to an advantageous behavior (modified Grætz) in which the development of boundary layers is truncated such that the long-time values of the transfer coefficients can remain substantially larger than in uniaxial cases. We reiterate though, that in many practical scenarios, the quantitative differences in rates of transfer between chaotic and nonchaotic flows may be small. Accurate models of these phenomena can be developed based on conventional ideas that date back more than a century. In other words, the presence of complex flows in the bulk does not require the use of an exotic theoretical framework in order to achieve predictive models and mechanistic insights.

References

1 Acharya, N., Sen, M., and Chang, H.C. (1992) Heat-transfer enhancement in coiled tubes by chaotic mixing. *International Journal of Heat and Mass Transfer*, **35** (10), 2475.

2 Peerhossaini, H., Castelain, C., and Leguer, Y. (1993) Heat exchanger design based on chaotic advection. *Experimental Thermal and Fluid Science*, **7**, 333.

3 Mokrani, A., Castelain, C., and Peerhossaini, H. (1997) The effects of chaotic advection on heat transfer. *International Journal of Heat and Mass Transfer*, **40** (13), 3089.

4 Chagny, C., Castelain, C., and Peerhossaini, H. (2000) Chaotic heat transfer for heat exchanger design and comparison with a regular regime for a large range of Reynolds numbers. *Applied Thermal Engineering*, **20**, 1615.

5 Ferrigno, R., Stroock, A.D., Clark, T.D., Mayer, M., and Whitesides, G.M. (2002) Membraneless vanadium redox fuel cell using laminar flow. *Journal of the American Chemical Society*, **124** (44), 12930.

6 Cohen, J.L., Volpe, D.J., Westly, D.A., Pechenik, A., and Abruna, H.D. (2005) A dual electrolyte H_2/O_2 planar membraneless microchannel fuel cell system with open circuit potentials in excess of 1.4V. *Langmuir*, **21** (8), 3544.

7 Shrivastava, A., Kumar, S., and Cussler, E.L. (2008) Predicting the effect of membrane spacers on mass transfer. *Journal of Membrane Science*, **323**, 247.

8 Brody, J.P. and Yager, P. (1997) Diffusion-based extraction in a microfabricated device. *Sensors and Actuators A-Physical*, **58**, 13.

9 Kenis, P.J.A., Ismagilov, R.F., and Whitesides, G.M. (1999) Microfabrication inside capillaries using multiphase laminar flow patterning. *Science*, **285**, 83.

10 Kenis, P.J.A., Ismagilov, R.F., Takayama, S., Whitesides, G.M., Li, S.L., and White, H.S. (2000) Fabrication inside microchannels using fluid flow. *Accounts of Chemical Research*, **33** (12), 841.

11 Kamholz, A.E., Weigl, B.H., Finlayson, B.A., and Yager, P. (1999) Quantitative analysis of molecular interaction in a microfluidic channel: The T-sensor. *Analytical Chemistry*, **71** (23), 5340.

12 Vijayendran, R.A., Motsegood, K.M., Beebe, D.J., and Leckband, D.E. (2003) Evaluation of a three-dimensional micromixer in a surface-based biosensor. *Langmuir*, **19** (5), 1824.

13 Golden, J.P., Floyd-Smith, T.M., Mott, D.R., and Ligler, F.S. (2007) Target delivery in a microfluidic immunosensor. *Biosens. Bioelectron.*, **22**, 2763.

14 Foley, J.O., Mashadi-Hossein, A., Fu, E., Finlayson, B.A., and Yager, P. (2008) Experimental and model investigation of the time-dependent 2-dimensional distribution of binding in a herringbone microchannel. *Lab on a Chip*, **8**, 557.

15 Squires, T.M., Messinger, R.J., and Manalis, S.R. (2008) Making it stick: Convection, reaction and diffusion in surface-based biosensors. *Nature Biotechnology*, **26**, 417.

16 Kirtland, J.D., McGraw, G.J., and Stroock, A.D. (2006) Mass transfer to reactive boundaries from steady three-dimensional flows in microchannels. *Physics of Fluids*, 18, 073602.

17 Cohen, J.L., Westly, D.A., Pechenik, A., and Abruna, H.D. (2005) Fabrication and preliminary testing of a planar membraneless microchannel fuel cell. *Journal of Power Sources*, **139** (1–2), 96.

18 Kirtland, J.D. (2009) *Interfacial Mass Transfer in Microfluidic Systems: Existence and Persistence of the Modified Graetz Behavior*, Cornell University, Ithaca, NY

19 Bird, R.B., Stewart, W.E., and Lightfoot, E.N. (2002) *Transport Phenomena*, 2nd edn, John Wiley & Sons, Danvers.

20 Graetz, L. (1885) Ueber die Wärmeleitungsfähigkeit von Flüssigkeiten. *Annalen der Physik*, **13**, 337.

21 Stroock, A.D., Dertinger, S.K.W., Ajdari, A., Mezic, I., Stone, H.A., and Whitesides, G.M. (2002) Chaotic mixer for microchannels. *Science*, **295** (5555), 647.

22 Stroock, A.D., Dertinger, S.K., Whitesides, G.M., and Ajdari, A. (2002) Patterning flows using grooved surfaces. *Analytical Chemistry*, **74** (20), 5306.

23 Stroock, A.D. and McGraw, G.J. (2004) Investigation of the staggered herringbone mixer with a simple analytical model. *Philosophical Transactions of the Royal Society of London Series a-Mathematical Physical and Engineering Sciences*, **362** (1818), 971.

24 Ottino, J.M. (1989) *The Kinematics of Mixing: Stretching, Chaos, and Transport*, Cambridge University Press, Cambridge.

25 Mary, P., Studer, V., and Tabeling, P. (2008) Microfluidic droplet-based liquid-liquid extraction. *Analytical Chemistry*, **80** (8), 2680.

26 Bard, A.J. and Faulkner, L.R. (2001) *Electrochemical Methods: Fundamental and Applications*, 2nd edn, John Wiley & Sons, New York.

27 Jones, S.W., Thomas, O.M., and Aref, H. (1989) Chaotic advection by laminar flow in a twisted pipe. *Journal of Fluid Mechanics*, **209**, 335.

28 Therriault, D., White, S.R., and Lewis, J.A. (2003) Chaotic mixing in three-dimensional microvascular networks fabricated by direct-write assembly. *Nature Materials*, **2**, 265.

29 Stroock, A.D., Ismagilov, R.F., Stone, H.A., and Whitesides, G.M. (2003) Fluidic ratchet based on Marangoni–Benard convection. *Langmuir*, **19** (10), 4358.

30 Bryden, M.D. and Brenner, H. (1996) Effect of laminar chaos on reaction and dispersion in eccentric annular flow. *Journal of Fluid Mechanics*, **325**, 219.

31 Ghosh, S., Leonard, A., and Wiggins, S. (1998) Diffusion of a passive scalar from a no-slip boundary into a two-dimensional chaotic advection field. *Journal of Fluid Mechanics*, **372**, 119.

32 Kirtland, J.D., Siegel, C.R., and Stroock, A.D. (2009) Interfacial mass transport in

steady three-dimensional flows in microchannels. *New Journal of Physics*, 11, 075028.

33 John, B.S., Escobedo, F.A., and Stroock, A.D. (2004) Cubatic liquid-crystalline behavior in a system of hard cuboids. *Journal of Chemical Physics*, **120** (19), 9383.

34 Ganesan, V., Bryden, M.D., and Brenner, H. (1997) Chaotic heat transfer enhancement in rotating eccentric annular-flow systems. *Physics of Fluids*, **9** (5), 1296.

35 Bryden, M.D. and Brenner, H. (1999) Mass-transfer enhancement via chaotic laminar flow within a droplet. *Journal of Fluid Mechanics*, **379**, 319.

36 Chang, H.C. and Sen, M. (1994) Application of chaotic advection to heat-transfer. *Chaos Solitons & Fractals*, **4** (6), 955.

37 Danckwerts, P.V. (1951) Significance of liquid-film coefficients in gas adsorption. *Industrial and Engineering Chemistry*, **43** (6), 1460.

38 Leveque, A. (1928) Les lois de la transmission de chaleur par convection. *Annales des Mines, Memoires*, **12 & 13**, 201.

5
The Effects of Laminar Mixing on Reaction Fronts and Patterns
Tom Solomon

5.1
Introduction

A wide variety of dynamical processes are influenced by some sort of reaction. In its most general sense, a "reaction" is when some species A changes into some other species B during a process, possibly via an interaction with other species. The most obvious example is a chemical reaction, when two or more reactants interact to form a third product, for example, A + B → C or A + B → 2A. Combustion is an example of a chemical reaction, either in a controlled environment such as an engine or in an uncontrolled environment such as a forest fire. But reaction dynamics are found in other fields of science and engineering. In biology, for example, life-cycle processes can be treated as reactions with different states when an animal is born, grows, and dies. Contraction of a disease can be treated as a reaction with at least two distinct states (healthy and sick). And various processes within living organisms can be treated as reactions; for example, waves of electrical activity in the heart are treated as reactions. Other examples of reactions include phase transitions where, for example, a solid (state "A") turns into a liquid ("B").

Reaction dynamics are heavily dependent on mixing in the system. For a chemical reaction in a flask, if the reactants are well-mixed, it can often be assumed that the reaction occurs everywhere in the system simultaneously. However, real systems are often spatially extended, and there are typically variations in concentration of the species in the system. The reactants need to be brought into proximity with each other to undergo a reaction; furthermore, the products of the reaction can interact with the remaining reactants. In the absence of any fluids flows, mixing is entirely via molecular diffusion. If there are fluid flows in the system, mixing is significantly enhanced over that due to molecular diffusion alone; consequently, fluid flows typically have a significant effect on the behavior of the system. If the flows are strongly turbulent, the mixing may be sufficiently strong such that the spatial extent of the system is not an issue (i.e., similar to a reaction in a well-stirred flask, the reaction occurs roughly simultaneously everywhere). However, it is often not possible to get strong enough turbulent mixing.

Transport and Mixing in Laminar Flows: From Microfluidics to Oceanic Currents,
First Edition. Edited by Roman Grigoriev.
© 2012 Wiley-VCH Verlag GmbH & Co. KGaA. Published 2012 by Wiley-VCH Verlag GmbH & Co. KGaA.

The strength of a fluid flow is typically characterized by the Reynolds number $Re = UL/\nu$, where U and L are the characteristic velocity and length scale of the flow, respectively, and ν is the kinematic viscosity of the fluid. Turbulence requires a large Re, which is difficult for highly viscous flows or flows with a small characteristic length scale. The latter condition is particularly relevant for cellular-scale processes in biological organisms and for microfluidic devices, both of which have length scales that are often smaller than a micron. For small Re, the flow is laminar; in this situation, it is rarely the case that a reaction can be assumed to happen simultaneously throughout a system. In this case, the evolving reaction forms spatial patterns that are dependent on the combination of local reaction dynamics and longer range mixing between different parts of the system.

The simplest types of spatially extended reacting systems are *reaction–diffusion* systems which, by definition, have no fluid flows. Mixing is entirely due to molecular diffusion, a process which is inherently local in nature with each part of the fluid interacting only with the surrounding fluid. Reaction–diffusion systems have been studied extensively for several decades. For oscillating reactions (e.g., the oscillating Belousov–Zhabotinsky chemical reaction or life-cycle processes for living organisms) or for "excitable" reactions (one-time reactions that can reset, such as the excitable Belousov–Zhabotinsky reaction, a disease from which someone can recover, electrical waves in the heart or brain, or a forest fire with trees that can regrow), it is common for rotating spiral and expanding target patterns to form. For "one-off" ("burn-type") reactions that do not reset (e.g., combustion of a fuel, solidification of a cooling melt, or the propagation of a deadly disease), it is typical for fronts to develop which propagate across the system with a well-defined front velocity that depends on the molecular diffusivity and reaction rates of the interacting species.

The addition of a fluid flow dramatically changes the pattern-formation and front propagation behavior [1]. This is relevant to a wide range of reacting systems, including ecosystems in oceanic flows, forest fires in the presence of wind, and chemical and biological processing systems that are subject to fluid flows. There has been a significant amount of previous research into the effects of *turbulent* flows on reaction dynamics (e.g., in turbines and in premixed burners). Surprisingly, however, there has not been much research until recently on the effects of *laminar* fluid flows on reaction dynamics. Laminar flows are relevant for a wide range of reacting systems, including microfluidic reactors (so-called laboratories on a chip) and biological processes in cellular and embryonic systems, all of which occur on small-enough scales such that Reynolds numbers are too small to sustain turbulent flows.

This chapter discusses recent research on the effects of laminar fluid mixing on reaction dynamics. We review both theoretical and experimental studies, and discuss areas that are currently being studied and for which significant additional research is needed. In Section 5.2, we present background material on relevant issues in fluid mixing, along with previous results in reaction–diffusion systems. Section 5.3 covers some of the basic principles common to *advection–reaction–diffusion* (ARD) systems, that is, reacting systems with fluid mixing. Section 5.4 discusses recent studies that examine how laminar fluid mixing affects local pattern formation in ARD systems.

Section 5.5 discusses synchronization of processes in extended fluid systems and how laminar mixing relates to that synchronization. Section 5.6 discusses the issue of front propagation and how laminar mixing affects the process. Section 5.7 summarizes and discusses future areas of investigation.

5.2 Background

5.2.1 Laminar mixing – the Advection–Diffusion Equation

Mixing is governed by the interplay between advection of impurities along streamlines as they are carried by a fluid flow and diffusion of the impurities between streamlines in the flow. Quantitatively, the process is governed by the advection–diffusion equation (also known as the *convection–diffusion equation*):

$$\partial c/\partial t + \vec{v} \bullet \vec{\nabla} c = D\nabla^2 c$$

where $c(\vec{r}, t)$ is the concentration field for the impurity, $\vec{v}(\vec{r}, t)$ is the velocity field, and D is the molecular diffusion coefficient for the impurity in the background fluid. Conceptually, the concentration of the impurity at a location changes either if the flow carries a region of lower or higher concentration to the point of interest, or if there is diffusion of impurity into our out of the point of interest from the neighboring regions. In nondimensional form, the advection–diffusion equation is written as

$$\partial c/\partial t = \tfrac{1}{Pe}\nabla^2 c - \vec{v} \bullet \vec{\nabla} c$$

where all the quantities have been nondimensionalized by the characteristic velocity U and length L scales of the flow, and the *Peclet number* is defined as $Pe = UL/D$. Conceptually, the Peclet number characterizes the relative strength of the advection and diffusion terms. Mixing with $Pe \gg 1$ is dominated by advection, whereas $Pe \ll 1$ corresponds to diffusion-dominated mixing.

5.2.2 Short-Range Mixing

Several techniques have been developed to characterize local mixing in fluid flows. Many of these techniques involve quantification of the manner in which the flow stretches fluid elements. Quantitatively, stretching is characterized by a positive finite-time Lyapunov exponent (FTLE) $\lambda(t)$ defined by

$$\lambda(t) = \frac{1}{t}\ln\left(\frac{\Delta r(t)}{\Delta r_0}\right)$$

where $\Delta r(t)$ is the separation of two passive tracers after a time t, and Δr_0 is the initial separation of those two tracers (assumed to be much smaller than typical length scales of the flow). If mixing is chaotic, then a positive Lyapunov exponent can be found even in the infinite-time limit. The FTLE can be represented as a field, showing visually the regions of maximal stretching in a flow.

Stretching of fluid elements in itself is not the prime contributor to mixing; rather, compression of streaks of impurity is the key. However, since most laminar flows are incompressible, significant stretching of fluid elements (and large positive Lyapunov exponents) is associated with significant compression of the same fluid elements in an orthogonal direction, characterized by large negative Lyapunov exponents. Compression of fluid elements results in thinning of tendrils of impurity in the flow. Ultimately, complete mixing occurs as a two-stage process: advection of the impurity in the flow stretches (and thins) and folds the impurity repeatedly until the structures are thin enough so that molecular diffusion can finish the job.

Recently, there has been a tremendous interest in what are referred to as *Lagrangian coherent structures* (LCSs) as a method for identifying regions of significant mixing in a fluid flow. These structures are equivalent to *unstable manifolds* of fixed points in a fluid flow, as shown in Figure 5.1. The unstable manifold of a fixed point is defined as the future location of a swarm of tracer particles that start out infinitesimally close to the fixed point. For a time-independent flow (Figure 5.1a), the unstable manifold of a fixed point is usually a simple curve that separates different mixing regions in the flow; that is, it is frequently equivalent to the separatrix. If the flow is time-periodic (Figure 5.1b), the manifold undulates, forming a pattern of folds that become increasingly complicated in time and distance from the fixed point. A stable manifold can also be defined for a fixed point, either as the set of points that in the distant future will end up arbitrarily close to the fixed point, or as the set-up points which would denote the unstable manifold of that fixed point of time were reversed.

From a perspective of short-range mixing, the unstable manifold is the region of the flow where stretching is maximized; consequently, compression in the orthogonal direction is also maximized and mixing is strong and efficient. These manifold techniques were pioneered a couple of decades ago for time-independent and time-periodic flows. Recently, research groups have successfully extended these ideas to

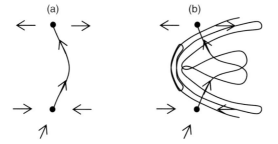

Figure 5.1 Stable and unstable manifolds of two hyperbolic fixed points in a flow. (a) Time-independent flow; the unstable manifold of the lower fixed point is the same as the stable manifold of the upper fixed point. (b) Time-periodic flow; the unstable manifold of the bottom fixed point undulates, as does the stable manifold of the upper fixed point.

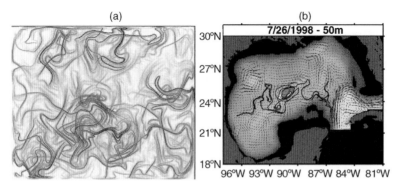

Figure 5.2 Stable and unstable manifolds in more complicated flows. (a) Manifolds from weakly turbulent magnetically forced flow (from Greg Voth and Jerry Gollub; see Ref. [2]). (b) Manifolds and hyperbolic trajectory of Eddy Fourchon in the Gulf of Mexico (from Ref. [3]).

enable the experimental identification of both stable and unstable manifolds to turbulent fluid flows [2] (Figure 5.2a). These approaches have been used very successfully to analyze mixing in flows in the oceans and in the Gulf of Mexico [3, 4].

5.2.3
Long-Range Transport of Impurities

Stable and unstable manifolds can also be used to quantify long-range transport in extended fluid flows; specifically, over distances larger than characteristic length scales in the flow. A paradigmatic flow for these studies is the alternating vortex chain (Figure 5.3), a flow that is common in nature, for example, in flows with thermal convection, sheared flows, and cloud streets. If we approximate the flow as two-dimensional (2D) or consider a 2D cross-section of a 3D flow, each vortex is a separate mixing region, and long-range transport is achieved only if there is a mechanism for tracers to cross the separatrices between adjacent vortices. For a time-independent flow, the separatrices are manifolds for the fixed points at the vortex corners, and transport between vortices occurs only via molecular diffusion of impurity across these separatrices. If the flow is time-periodic, however, the folded stable and unstable manifolds intersect to form a pattern of lobes (Figure 5.4) that provides an advective mechanism that enables impurities to cross from one vortex to the next. In Figure 5.4, impurities in the bottom filled-in lobe end up in the next lobe (above the first and to the left of the separatrix) one period later, then the next lobe (near the top)

Figure 5.3 Sketch of alternating vortex chain. In the studies presented in this chapter, the vortex chain can move laterally, either oscillating periodically with a maximum lateral speed v_o, moving with a constant drift velocity v_d, or moving with a combination of oscillatory and drifting motion.

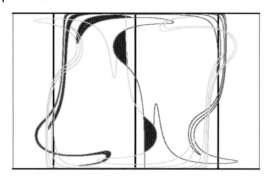

Figure 5.4 Stable and unstable manifolds of the hyperbolic fixed points for the alternating vortex chain. The straight vertical lines are the separatrices between adjacent vortices, and are also the manifolds for the time-independent case with no lateral oscillations. The complicated (stretched and folded) manifolds correspond to time-periodic, lateral oscillations of the vortex chain. Some of the lobes formed from the intersections of these manifolds are shaded in.

one period after that, and so on. Similarly, impurities in the bottom, clear lobe end up in the middle lobe (straddling the separatrix) one period later, then the top, clear lobe, and so on. This "turnstile" mechanism results in transport of impurities between adjacent vortices, with the transported amount being determined by the area of the lobes.

To first order, this lobe mechanism of transport results in long-range transport that is essentially diffusive with an impurity distribution whose variance grows linearly in time: $\langle x^2(t) \rangle = 2D^*t$, where D^* is the effective (enhanced) diffusion coefficient. Experiments have been done [5, 6] to test these ideas quantitatively; the results of the experiments verify this lobe picture of long-range transport, along with the typically diffusive nature of the transport.

Long-range transport in an advection–diffusion system is not necessarily diffusive. More generally, the variance grows as a power law in time: $\langle x^2(t) \rangle \sim t^\gamma$. If $\gamma = 1$, then the transport is diffusive. If $\gamma \neq 1$, then the transport is deemed "anomalous" with $\gamma < 1$ corresponding to *subdiffusive* transport and $\gamma > 1$ corresponding to *superdiffusive* transport. Superdiffusive transport is associated with trajectories called *Lévy flights* [7–9] where tracers in the flow undergo sporadic "jumps" whose lengths follow a power-law probability distribution: $P(L) \sim L^{-\mu}$, where $\mu < 3$. This is a probability distribution whose second moment diverges: $\langle L^2 \rangle = \infty$. Lévy flights and superdiffusion were first seen experimentally in a chain of time-periodic *corotating* vortices [10], but they can also be found in the time-periodic counter-rotating vortex chain if either the oscillation frequency is resonant with a typical circulation time [11] or if a uniform wind is added to the flow [12] with velocity magnitude $W > v_{osc}$, where v_{osc} is the maximum lateral oscillation velocity of the vortex chain.

Since the first experiments in the early 1990s that showed Lévy flights and superdiffusion, Lévy flights have been discussed for a wide range of systems, including the motion of people in society [13] and the foraging patterns of various animals [14, 15]. Consequently, superdiffusion is most likely relevant to a range of real systems.

5.2.4
Nonlinear Reactions

In the mid-1950s, Boris Belousov discovered a chemical reaction that approaches equilibrium in a long, cyclical process in which the pH of the solution oscillates for up to several hours. When tagged with an indicator, the solution changes color, alternating between two different colors through the life of the reaction. The same effect was rediscovered a few years later by Anatol Zhabotinsky. The reaction and its variants – which are now referred to broadly as the *Belousov–Zhabotinsky* (BZ) reaction – have been studied extensively for the past few decades as a paradigm for nonlinear reactions [16–18].

Two different regimes are common for BZ and similar reactions, depending on the relative concentrations of the reactants: (1) an oscillatory regime in which the pH spontaneously oscillates periodically (or almost periodically) for many oscillation periods before the reaction finally reaches equilibrium; and (2) an excitable regime in which the reaction requires some sort of trigger to change its pH (and its color if an indicator is used), but then returns to its initial pH and color.

The BZ reaction has also been manipulated to produce chaotic time dependence, if configured in a continuously stirred tank reactor (CSTR) where new chemical reactants are continuously fed into the system.

5.2.5
Reaction–Diffusion Systems

When nonlinear reactions are in an extended system, the interaction between local reaction dynamics and diffusive mixing results in a variety of spatial patterns and front-producing behavior. In the absence of a fluid flow, these systems are referred to generally as *reaction–diffusion* (RD) systems, described generically by the reaction–diffusion equation

$$\partial c_i/\partial t = D_i \nabla^2 c_i + f_i(c_1, c_2, c_3, \ldots)$$

Here, c_i is the concentration of one of the species in the system, D_i is the molecular diffusion coefficient for that species, and f_i is the reaction term, which depends not only on c_i but also on the other species in the system.

The behavior of fronts moving in a reaction–diffusion system has been well studied for reactions that are described [19, 20] by a complex function $f(c)$ where $f'' < 0$. In this case, the front propagates with a speed $v_{\rm rd} = \sqrt{2D/\tau_r}$ where D is the molecular diffusion coefficient for the relevant reactants and τ_r is the timescale for the reaction. This theory – referred to as *FKPP* theory since it was developed separately by Fisher [21] and by Kolmogorov et al. [22] – is well established, and applies to a wide range of reacting systems, not only chemical but biological and physical as well, as long as a diffusivity and reaction time scale can be identified.

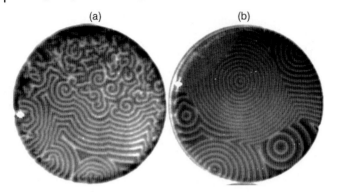

Figure 5.5 (a) Spiral and (b) target patterns in the excitable Belousov–Zhabotinsky chemical reaction in a Petri dish with no imposed fluid flow.

Reaction–diffusion systems are also well known for their ability to form patterns [16]. The BZ reaction in its excitable regime readily forms patterns composed of target and spiral patterns, as shown in Figure 5.5. Conceptually, these patterns – referred to as *trigger waves* – form due to the ability of a triggered excitable reaction to reset back to (roughly) its original state, after which it can be retriggered. A rotating spiral is an example of a self-sustaining pattern for an excitable system. This kind of self-sustaining pattern formation behavior appears even if the reaction is oscillatory. Initially, the entire reaction oscillates together, but over time, the oscillations desynchronize, forming regions of *phase waves*. In some of these regions, seeds of trigger waves form, and these trigger waves grow in extent into the phase wave regions.

Pattern formation in the BZ chemical reaction in the RD limit has been studied extensively, predominately because the behavior seen in this system is typical of pattern formation seen in a wide range of RD systems. Examples – shown in Figure 5.6 – include waves of electrical activity in the heart [23] which are responsible for the heart's pace-maker, *spreading depression* in the visual cortex [24] which is responsible for the visual patterns seen in migraine headaches, and patterns formed in slime mold cultures [25] (*Dictyostelium discoideum*).

5.3
Advection–Reaction–Diffusion: General Principles

When there is both a reaction and a fluid flow, the system can be described in general by the ARD equation

$$\partial c_i/\partial t = -\vec{v} \cdot \vec{\nabla} c_i + D_i \nabla^2 c_i + f_i(c_1, c_2, c_3, \ldots)$$

which combines the advection–diffusion and reaction–diffusion equations. Conceptually, this equation states simply that the concentration of a species at a location changes due to, respectively, advection of that species from other locations due to a

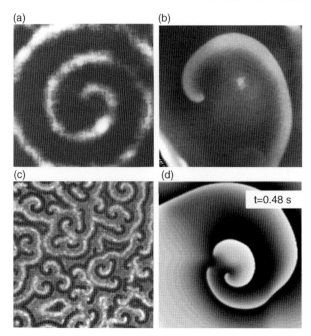

Figure 5.6 Reaction–diffusion patterns in other systems. (a) Developing frog embryos (David Clapham, May Foundation). (b) Spreading depression in the visual cortex (Stefan Muller, Univ. Magdeburg). (c) Slime mold cultures (Cornelis Weijer, University of Dundee). (d) Electrical waves in heart (Qu, Weiss and Garfinkel).

fluid flow ($-\vec{v}\bullet\vec{\nabla}c_i$), diffusion to and from neighboring regions ($D_i\nabla^2 c_i$), and reaction of c_i with other species that can change the concentration locally ($f_i(c_1, c_2, c_3, \ldots)$). Written in nondimensional form [26], the ARD equation becomes

$$\partial c_i/\partial t = -\vec{v}\bullet\vec{\nabla}c_i + Pe^{-1}\nabla^2 c_i + Da \cdot f_i(c_1, c_2, c_3, \ldots)$$

where we have introduced the *Damkohler number* $Da = L/U\tau_r$. The Damkohler number can be written as the ratio of the advective timescale τ_a to the reaction time scale τ_r: $Da = \tau_a/\tau_r$; consequently, a large Da corresponds to an ARD system in which the flow does not move the reactive species very far during a typical reaction time, whereas a small Da corresponds to an ARD system in which the reactive species are mixed significantly during typical reaction times.

The Damkohler number does not, however, take into account the role of diffusion in the ARD process, and the Peclet number (which weighs the relative strengths of advection and diffusion) does not account for the reaction. Another dimensionless number [27] that is appropriate for front behavior in ARD systems can be defined by the ratio of the characteristic advective velocity U to the speed v_{rd} at which a front propagates in the reaction–diffusion limit (i.e., $U \rightarrow 0$): $\mu = U/v_{rd}$.

There have been numerous studies of ARD systems with turbulent flows; in particular, the problem of turbulent combustion has received a lot of attention due to its clear applications in various fields of engineering. However, the behavior of ARD systems in laminar flows has only recently received attention, most of which is theoretical.

5.4
Local Behavior of ARD Systems

The tools used to describe chaotic mixing have been used successfully to characterize the patterns that form in ARD systems. Locally, chemical patterns tend to follow the mixing structures in the flow. Experiments in the early 1990s [28] measured regions of maximum chemical activity in one-time reactions and found that these regions matched up well with the regions of maximum stretching of impurities in the flow. More recent experiments in magnetically driven vortex flows [29] also used stretching fields to characterize regions of significant chemical activity in one-time reactions.

The use of stretching fields to characterize chemical activity makes sense, considering that the regions where fluid elements are most significantly stretched are also regions of maximal compression in orthogonal directions and are therefore regions where molecular diffusion can mix impurities most efficiently. Note that this is effectively equivalent to using fields of finite-time Lyapunov exponents, which is effectively the logarithm of the stretching field, normalized by the time span.

Theoretical studies [1] during the first decade in the 2000s described this phenomenon in terms of unstable manifolds of fixed points in the flow. Theory and modeling have been done for burn-type reactions in a blinking vortex flow (Figure 5.7) with source and sink terms. Mixing in this flow is chaotic, similar to that

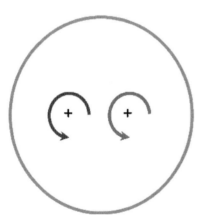

Figure 5.7 Blinking vortex flow. The flow has periodic time-dependence. During half of the period, fluid circulates around the left point vortex; during the other half of the period, fluid circulates around the right point vortex. The flow continues to "blink" periodically between these two vortices. The source-sink version of this flow has sources and sinks of fluid at the vortex centers.

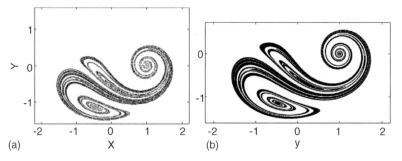

Figure 5.8 Numerical unstable manifold (a) and pattern for autocatalytic reaction (b) for blinking vortex–sink system. (T. Tél et al., Ref. [1]).

in Hassan Aref's original blinking vortex flow model of 1984, but the addition of a source and sink makes this an "open" flow in which the reactants are replenished. So, instead of the region of "burned" reactants growing indefinitely until it covers the entire system, there is ultimately a balance condition and a steady-state pattern forms around the unstable manifolds of the flow (Figure 5.8).

Similar behavior is observed for "closed" flows with oscillating reactions [30]. Since the reaction oscillates rather than burning across the entire system, it is not necessary to replenish the reactants to avoid having the burned front growing in area indefinitely. Experiments were done in a blinking vortex flow (without sources and sinks) and the patterns that formed in these experiments agreed very well with patterns obtained from *mixing fields* of the flow which are basically stretching fields but chosen with a timescale equal to the time for the BZ pattern to decorrelate in the absence of any fluid flows. Examples from these experiments are shown in Figure 5.9.

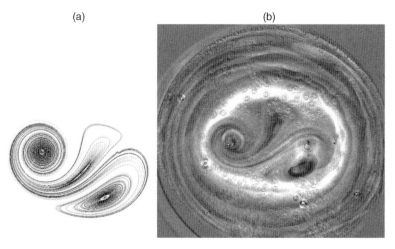

Figure 5.9 (a) Mixing field for blinking vortex flow. (b) Pattern formed by oscillating Belousov–Zhabotinsky reaction in blinking vortex flow (Ref. [30]).

Finally, the behavior discussed in this section for chemical reactions in both open and closed flows applies to other, nonchemical systems. In particular, several studies [31, 32] have investigated "blooms" of algae and phytoplankton in the oceans and Gulf of Mexico, using the same analytical techniques involving manifolds for mixing in these bodies of water.

Another issue involving local pattern formation in ARD systems is that of *extinction*; that is, when the mixing is strong enough so that the reaction is terminated (like a flame being blown out). In fact, some reactions are bistable where the reaction either spreads and eventually covers the entire system (space-filling) or is completely extinguished by the flow [33].

5.5
Synchronization of Oscillating Reactions

In a fluid system, mixing is the key to understanding when time-varying processes synchronize. In the studies of pattern formation for oscillating reactions discussed in the previous section, the thickness of the oscillating structures increases with the mixing efficiency of the flow. In the limit of perfect mixing, the thickness of the structures diverges to the length of the entire system; in this limit, the reaction is synchronized throughout the system.

A few studies have investigated how mixing contributes to synchronization in fluid networks, where individual nodes can be identified and where the mixing between these nodes is controlled. In one series of studies [34], the chaotic mixing in the system is analyzed theoretically as a global coupling between different oscillators. Regimes of complete synchronization are found, as well as regimes of "oscillator death" when the coupling due to mixing ends the reaction entirely (similar to extinction, discussed in the previous section).

Experiments have been done on coupling of individual oscillators. In one study [35], an array of up to 64 Ni electrodes was immersed in an electrochemical apparatus. Coupling in this experiment is achieved electronically, however, rather than via fluid mixing. These experiments demonstrated that sufficient coupling could lead to global synchronization or oscillator death, depending on the circumstances.

In an extended system, the type of long-range transport affects the manner in which different parts of the system are coupled. As discussed in Section 5.2.3, diffusive transport is associated with fluid elements that undergo random-like walks between neighboring regions in the flow, whereas superdiffusive transport is associated with Lévy flight trajectories where fluid elements can travel large distances in a short period of time. From a network perspective, a system with diffusive transport is akin to a network with nearest-neighbor coupling (Figure 5.10a), whereas a system with superdiffusive transport is analogous to a network with long-range connections (Figure 5.10c). This is reminiscent of the studies done at the end of the 1990s on network models and, in particular, on the small-world network models [36] that use long-range "short-cut" connections to enhance dramatically the connectivity of a network.

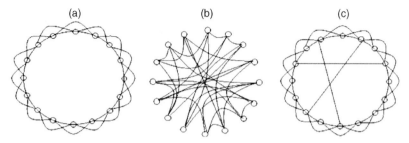

Figure 5.10 Different models for connected networks. (a) Nearest-neighbor model where each node of the network is connected with its two nearest neighbors. (b) Random model where nodes are connected in a random pattern. (c) "Small-World" network which is predominately a nearest-neighbor model, but with a few, long-distance "short-cut" connections.

Two experimental programs have studied the effects of different types of long-range transport on the behavior of a network of chemical oscillators. In one study [37], instead of using a fluid flow, the long-range "transport" is simulated experimentally by using a photo-sensitive (Ruthenium-catalyzed) version of the BZ reaction and a feedback mechanism using imaging and an LCD projector. Long-distance coupling is achieved by measuring fluctuations in the reaction at a particular location and then adjusting the projector intensity at another location in response to that fluctuation. If the long-range feedback is implemented with a power-law relation (i.e., describing coupling amplitude versus distance, similar to Lévy flights) then long-range synchronization of the reaction dynamics is found.

Another concurrent study [12] studied the effects of fluid mixing on long-range synchronization more directly. The flow studied was the oscillating/drifting vortex chain described in Section 5.2.3 and shown in Figure 5.3. The fluid is composed of the chemicals for the BZ reaction, and each vortex acts as an individual chemical oscillator that is coupled to other nodes in the network via fluid mixing. In the absence of any oscillation or drift of the vortices, each vortex is essentially isolated from its neighbors except via molecular diffusion; consequently, the BZ oscillations occurring within the vortices become completely desynchronized. If there is lateral motion of the vortices and if the drift velocity (equivalent to the speed of an imposed wind in a codrifting reference frame) v_d is less than the maximum lateral oscillation speed v_o of the vortices, then the transport is diffusive and the coupling is effectively nearest-neighbor. In this situation, the system spontaneously forms traveling-phase waves with a complicated (possibly chaotic) time evolution (Figure 5.11a). On the other hand, if $v_d > v_o$, the transport is superdiffusive (with Lévy flights) and there is significant long-range coupling between the reactions oscillating within the vortices (Figure 5.11b and c). In this situation, the oscillations within the vortices rapidly synchronize globally. The result is clear: the type of long-range fluid mixing has a significant effect on collective behavior of a fluid network of oscillating reactions.

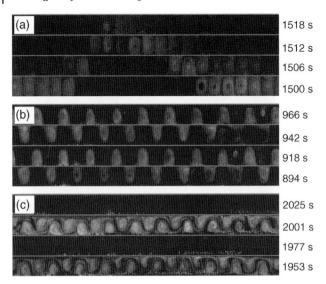

Figure 5.11 Sequences of images of oscillating BZ reaction in an annular chain of oscillating/drifting vortices. The annulus has been "de-curled" to show the vortices in a strip. (a) Periodic lateral oscillations. Traveling waves form that propagate around the annulus in an unpredictable manner. (b) Periodic lateral oscillations with a superposed DC drift; mixing is superdiffusive, coupling every second vortex. Oscillations of the BZ reaction in the vortices synchronize in alternating (odd or even) vortices. (c) Oscillating/drifting time dependence with superdiffusive mixing that connects all the vortices. The BZ oscillations are globally synchronized. (From Ref. [12]).

The issue of the effects of fluid mixing on synchronization is still a very open issue with many unanswered questions. Theoretical modeling of the drifting/oscillating vortex chain experiments has not been done yet. There are differences between the simple Watt/Strogatz small-world network model [36] and the coupling due to superdiffusion and Lévy flights in an extended fluid system. First, coupling due to superdiffusion is not only between random vortices; rather, every vortex is coupled to every other vortex in the system with a coupling strength that decays as a power law with distance. Second, the duration of the flights has to be taken into account; fluid elements in a flight take a finite time to complete the excursion, and if that time exceeds typical correlation times for the reaction, that could affect the ability of superdiffusion to synchronize an extended fluid network.

Another open question is how applicable these ideas are to a range of oscillating systems occurring either in fluid flows or as part of a "flowing" system. For instance, networks of people in a moving population may be considered – for a sufficiently large number of people – as a flow, in which case it might be instructive to model dynamical processes affecting moving populations as a continuous ARD system.

5.6
Front Propagation in ARD Systems

There have been very few theoretical studies of how front propagation is affected by the presence of a fluid flow. In the case where the enhanced transport is diffusive, the problem for front propagation seems almost obvious: simply take the FKPP prediction $v_{rd} = \sqrt{2D/\tau_r}$ and replace the molecular diffusion coefficient with the enhanced diffusivity D^* describing transport in the flow. However, as we discuss below, there are experiments for which this approach clearly does not work. There have been several theoretical studies that extend FKPP theory to cases with superdiffusive transport [38], most using techniques derived from fractional calculus, which is a mathematical language that is well-suited for incorporating long-distance interactions produced by Lévy flights. However, the fact that an FKPP approach does not adequately describe front propagation in ARD systems with enhanced diffusion raises questions about whether a modified version of FKPP theory will work for superdiffusive cases. (Experimental data for the superdiffusive case is still lacking.)

The main issue when dealing with front propagation in ARD systems appears to be the role of coherent flow structures on the front-propagation phenomena. In particular, vortices play a significant role in the process beyond the role expected simply by their effect on the enhanced diffusivity. The first indications of difficulties with an FKPP approach to front propagation in ARD systems were provided by a series of numerical studies [19, 20] and experimental studies [39] that studied front propagation in the oscillating vortex chain flow discussed in Section 5.2.3 (Figure 5.3). Numerical and experimental sequences of images of a front in this flow are shown in Figure 5.12. The fronts are observed to *mode-lock* to the external, time-periodic, lateral oscillations of the vortex chain, moving an integer number N of unit cells λ of the flow (where one unit cell λ is two vortex widths) in an integer number M of oscillation periods. Figure 5.12a and b shows an example of $(N,M) = (1,1)$ mode-locking where the front moves 1 unit cell (two vortex widths) each drive period, and Figure 5.12c and d shows an example with (1,2) mode-locking where the front moves 1 unit cell in 2 drive periods (and 1 vortex every period, although the front flips each period).

The front velocity when mode-locked depends on the combination (N,M), the vortex width and the drive period T or frequency f: $v_f = (N\lambda)/(MT) = (N/M)\lambda f$. The (enhanced) diffusivity of the chemical species and the reaction time scales do not directly affect the front-propagation speed. A plot of experimental front speeds is shown in Figure 5.13, along with the predictions for mode-locked fronts. There are no fitted parameters in this figure – the experimental data agree almost perfectly with the mode-locking predictions.

The horizontal line in Figure 5.13 shows the front speed in the absence of any lateral oscillations of the vortex chain. The fact that some of the front speeds with the time-periodic oscillations are actually *slower* than the front speed for the time-independent flow is the clearest evidence of the inapplicability of an enhanced-FKPP approach, since the effective diffusivity for any time-periodic flow (with lateral oscillations of the vortex chain) is significantly larger than that for the

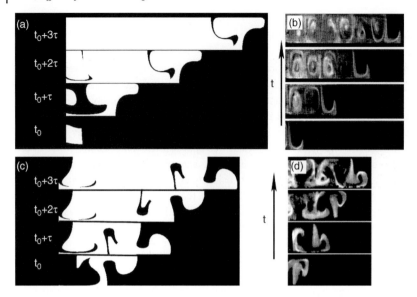

Figure 5.12 Mode-locking for front propagation in an oscillating chain of vortices. (a) Simulation and (b) experiments showing (1,1) mode-locking where the front moves one wavelength (two vortices) each drive period. (c) Simulation and (d) experiments showing (1,2) mode-locking where the front moves one wavelength in two drive periods (or one vortex every period).

time-independent case (no lateral oscillations), so a larger front speed would be expected from an FKPP approach based on the enhanced diffusivity.

The key appears to be the behavior of a reaction front when it encounters a vortex in the flow, especially if the vortex is moving. Experimental studies [27] have

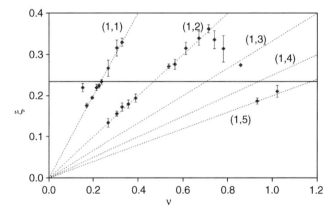

Figure 5.13 Experimentally determined reaction front velocities in oscillating vortex chain. The front velocities ξ are nondimensionalized by the maximum flow velocity U, and the frequency ν of oscillation is nondimensionalized by U and by the vortex width $d = \lambda/2$. The diagonal lines are the front speeds predicted for mode-locking (From Ref. [39]).

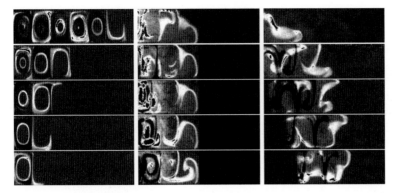

Figure 5.14 Reaction fronts in the presence of a leftward-directed wind imposed on the vortex chain. Time increases going upward in these sequences. If the wind speed W is smaller than the reaction–diffusion front speed v_{rd} (left sequence), then the front propagates to the right against the wind. For moderate wind speeds ($v_{rd} < W$, but W can be several times – and more than an order of magnitude – greater than v_{rd}), as in the middle sequence, the front remains "frozen" to the leading vortex, propagating neither forward nor backward. For large enough wind speeds (right sequence), the front is blown backward against the wind. (From Ref. [40]).

demonstrated that moving vortices pin and drag reaction fronts. Alternately, in a reference frame that is comoving with the vortices, the reaction front is "frozen" in the leading vortex in the face of an imposed wind, as shown in Figure 5.14. This behavior is seen not just for an ordered chain of vortices but also for random, vortex-dominated flows. Recent studies of the behavior of phytoplankton blooms in oceanic vortex flows [40] have also demonstrated the ability of moving vortices to pin and drag reacting species.

Ultimately, a general theory of front propagation in ARD systems is needed in which the role of coherent flow structures is taken into account. Studies are on-going about the possibility of a general approach based on pinning of fronts on vortices. Another approach being pursued is to try to extend the approaches used to describe chaotic fluid mixing to account for front propagation phenomena. In particular, it might be possible to extend manifold/lobe approach discussed in Sections 5.2.2 and 5.3 – that describes the motion of passive impurities in a flow – to account for the motion of a front in a flow, which moves by a combination of advection and reactive spreading.

The bottom line: a general theory to describe front propagation in ARD systems is still needed.

5.7
Additional Comments

The field of advection–reaction–diffusion for laminar fluid flows is still in its infancy. This is a field with a significant range of applications over a wide range of fields in science and engineering. At the smallest scales, laminar ARD behavior is relevant for microfluidic devices for which large Reynolds numbers are unachievable. Similarly, biological processes on a cellular or embryonic scale can be cast as ARD systems with laminar fluid flows. At everyday scales, an ARD analysis might shed light on plasma processes and might help in the development of safe nuclear fusion technologies. At larger scales, ARD phenomena have already been observed and explained for ecosystems in oceanic-scale flows. It has even been proposed that the motion of ignition fronts within a star during a supernova explosion might be modeled as an ARD phenomenon.

Finally, it is possible that even discrete systems can be modeled as ARD systems in the continuum limit. For example, instead of modeling the motion of a disease in a moving population using discrete network models, it might be more profitable to model the population as a flow and treat the spreading disease as a front that is propagating in this flow.

This work is supported by the US National Science Foundation, grants DMR-0703635 and DMR-0071771.

References

1 Tél, T., de Moura, A., Grebogi, C., and Károlyi, G. (2005) Chemical and biological activity in open flows: A dynamical system approach. *Physics Reports-Review Section of Physics Letters*, **413**, 91–196.

2 Voth, G.A., Haller, G.H., and Gollub, J.P. (2002) Experimental measurements of stretching fields in fluid mixing. *Physical Review Letters*, **88**, 254501.

3 Kirwan, A.D., Lipphardt, B.L.Jr., Poje, A.C., Kantha, L., and Zweng, M. (2007) 25 years of nonlinearity in oceanography from the Lagrangian perspective. In *Nonlinear Dynamics in Geosciences*, (eds A. Tsonis and J. B. Elsner), Springer, New York, pp. 177–196.

4 Olascoaga, M.J., Beron-Vera, F.J., Brand, L.E., and Kocak, H. (2008) Tracing the early development of harmful algal blooms on the West Florida Shelf with the aid of Lagrangian coherent structures. *Journal of Geophysical Research-Oceans*, **113**, C12041.

5 Solomon, T.H., Tomas, S., and Warner, J.L. (1996) The role of lobes in chaotic mixing of miscible and immiscible impurities. *Physical Review Letters*, **77**, 2682–2685.

6 Solomon, T.H., Tomas, S., and Warner, J.L. (1998) Chaotic mixing of immiscible and immiscible impurities in a two-dimensional flow. *Physics of Fluids*, **10**, 342–350.

7 Hughes, B.D., Shlesinger, M.F., and Montroll, E.W. (1981) *Proceedings of the National Academy of Sciences of the United States of America*, **78**, 3287.

8 Montroll, E.W. and Shlesinger, M.F. (1984) Nonequilibrium phenomena II: From stochastics to hydrodynamics, in *Studies in Statistical Mechanics*, vol. **II** (eds J.L. Lebowitz and E.W. Montroll), North-Holland, Amsterdam, p. 1.

9 Shlesinger, M.F., Zaslavsky, G.M., and Klafter, J. (1993) *Nature (London)*, **363**, 31.

10 Solomon, T.H., Weeks, Eric R., and Swinney, Harry L. (1993) Observation of anomalous diffusion and Lévy flights in a two-dimensional rotating flow. *Physical Review Letters*, **71**, 3975–3978.

11 Solomon, T.H., Lee, A.T., and Fogleman, M.A. (2001) Resonant flights and transient superdiffusion in a time-periodic, two-dimensional flow. *Physica D*, **157**, 40–53.

12 Paoletti, M.S., Nugent, C.R., and Solomon, T.H. (2006) Synchronization of oscillating reactions in an extended fluid system. *Physical Review Letters*, **96**, 124101.

13 Brockmann, D., Hufnagel, L., and Geisel, T. (2006) The scaling laws of human travel. *Nature*, **439**, 462–465.

14 Ramos-Fernández, G., Mateos, J.L., Miramontes, O., Cocho, G., Larralde, H., and Ayala-Orozco, B. (2004) Lévy walk patterns in the foraging movements of

spider monkeys (Ateles Geoffroyi). *Behavioral Ecology and Sociobiology*, **55**, 223–230.

15 Reynolds, A.M. (2006) Cooperative random Lévy flight searches and the flight patterns of honeybees. *Physics Letters A*, **354**, 384–388.

16 Grindrod, P. (1996) *The Theory and Applications of Reaction-Diffusion Equations: Patterns and Waves*, Clarendon Press, Oxford.

17 Field, R.J. and Burger, M. (1985) *Oscillations and Traveling Waves in Chemical Systems*, Wiley, New York.

18 Kuramoto, Y. (1984) *Chemical Oscillations, Waves and Turbulence*, Springer, Berlin.

19 Abel, M., Cencini, M., Vergni, D., and Vulpiani, A. (2002) *Chaos*, **12**, 481.

20 Cencini, M., Torcini, A., Vergni, D., and Vulpiani, A. (2003) *Physics of Fluids*, **15**, 679.

21 Fisher, R.A. (1937) *Proceedings of Annual Symposium Eugen Society*, **7**, 355.

22 Kolmogorov, A.N., Petrovskii, I.G., and Piskunov, N.S. (1937) *Moscow University Math Bull (English Translation)*, **1**, 1.

23 Qu, Z., Weiss, J.N., and Garfinkel, A. (1999) *American Journal of Physiology. Heart and Circulatory Physiology*, **276**, H269–H283.

24 Dahlem, M.A. and Müller, S.C. (2004) Reaction-diffusion waves in neuronal tissue and the window of cortical excitability. *Annalen der Physik*, **13**, 442–449.

25 Van Oss, C., Panfilov, A.V., Hogeweg, P., Siegert, F., and Weijer, C.J. (1996) Spatial pattern formation during aggregation of the slime mould Dictyostelium discoideum. *Journal of Theoretical Biology*, **181**, 203–213.

26 Neufeld, Z., Haynes, P.H., and Tel, T. (2002) Chaotic mixing induced transitions in reaction–diffusion systems. *Chaos*, **12**, 426.

27 Schwartz, M.E. and Solomon, T.H. (2008) Chemical reaction fronts in ordered and disordered cellular flows with opposing winds. *Physical Review Letters*, **100**, 028302.

28 Metcalfe, G. and Ottino, J.M. (1994) Autocatalytic processes in mixing flows. *Physical Review Letters*, **72**, 2875–2878.

29 Arratia, P.E. and Gollub, J.P. (2006) Predicting the progress of diffusively limited chemical reactions in the presence of chaotic advection. *Physical Review Letters*, **96**, 024501.

30 Nugent, C.R., Quarles, W.M., and Solomon, T.H. (2004) Experimental studies of pattern formation in a reaction–advection–diffusion system. *Physical Review Letters*, **93**, 218301.

31 Lehahn, Y., d'Ovidio, F., Lévy, M., and Heifetz, E. (2007) Stirring of the northeast Atlantic spring bloom: A Lagrangian analysis based on multisatellite data. *Journal of Geophysical Research*, **112**, C08005.

32 Hernández-Garcia, E. and López, C. (2004) Sustained plankton blooms under open chaotic flows. *Ecological Complexity*, **1**, 253–259.

33 Cox, S.R. (2006) Persistent localized states for a chaotically mixed bistable reaction. *Physical Review E*, **74**, 056206.

34 Zhou, C.S., Kurths, J., Neufeld, Z., and Kiss, I.Z. (2003) *Physical Review Letters*, **91**, 150601; Neufeld, Z., Kiss, I.Z., Zhou, C.S., and Kurths, J. (2003) *Physical Review Letters*, **91**, 084101.

35 Kiss, I.Z., Wang, W., and Hudson, J.L. (1999) Experiments on arrays of globally coupled periodic electrochemical oscillators. *The Journal of Physical Chemistry. B*, **103**, 11433–11444; Kiss, I.Z., Zhai, Y., and Hudson, J.L. (2002) Emerging coherence in a population of chemical oscillators. *Science*, **296**, 1676–1678.

36 Watts, D.J. and Strogatz, S.H. (1998) Collective dynamics of 'Small World' networks. *Nature*, **393**, 440–442.

37 Tinsley, M., Cui, J.X., Chirila, F.V., Taylor, A., Zhong, S., and Showalter, K. (2005) Spatiotemporal networks in addressable excitable media. *Physical Review Letters*, **95**, 038306.

38 Mancinelli, R., Vergni, D., and Vulpiani, A. (2002) *Europhysics Letters*, **60**, 532.; del Castillo-Negrete, D., Carreras, B.A., and Lynch, V.E. (2003) *Physical Review Letters*, **91**, 018302; Brockmann, D. and Hufnagel, L. (2007) *Physical Review Letters*, **98**, 178301.

39 Paoletti, M.S. and Solomon, T.H. (2005) Experimental studies of front propagation and mode-locking in an advection-reaction-diffusion system. *Europhysics*

Letters, **69**, 819–825; Paoletti, M.S. and Solomon, T.H. (2005) Front propagation and mode-locking in an advection–reaction–diffusion system. *Physical Review E*, **72**, 046204.

40 Sandulescu, M., Lopez, C.E., Hernandez-Garcia, E., and Feudel, U. (2007) Plankton blooms in vortices: The role of biological and hydrodynamics timescales. *Nonlin Processes Geophys*, **14**, 443–454.

6
Microfluidic Flows of Viscoelastic Fluids

Mónica S. N. Oliveira, Manuel A. Alves, and Fernando T. Pinho

6.1
Introduction

6.1.1
Objectives and Organization of the Chapter

In this chapter we provide an overview of viscoelastic fluid flow at the microscale. We briefly review the rheology of these nonlinear fluids and assess its implications on the flow behavior. In particular, we discuss the appearance of viscoelastic instabilities, which are seen to occur even under creeping flow conditions. The first type of instability changes the flow type from symmetric to asymmetric, while the flow remains steady. The second (and more frequent) type of instability, which sets in when elastic effects are enhanced, causes the flow to become unsteady varying in time periodically. This unsteadiness results in a nearly chaotic flow at higher flow rates, bringing about a significant improvement in mixing performance.

After a brief introduction to the theme of microfluidics, its basic principles, relevance and applications, this chapter is organized in five additional sections. Section 6.2 provides an overview of the problem of mixing at the microscale and of the current methods used to tackle this problem. Section 6.3 presents an introduction to non-Newtonian viscoelastic fluids describing their most relevant rheological properties. Section 6.4 presents the governing equations for Newtonian and non-Newtonian fluid flow, including the constitutive equations that describe the rheology of the fluids. Section 6.5 deals with passive mixing methods in viscoelastic fluid flows, whereas in Section 6.6 other forcing methods for promoting viscoelastic fluid flow at the microscale are briefly reviewed.

6.1.2
Microfluidics

6.1.2.1 Basic Principles, Relevance, and Applications
Microfluidics is a technological field that deals with the flow and handling of fluids in submillimeter-sized systems. Common microfluidic systems have features (typically

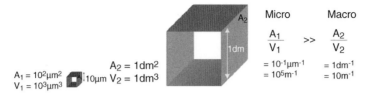

Figure 6.1 Surface-to-volume ratio: from macro- to microscale.

the channel width) with characteristic dimensions on the order of 10s to 100s of microns [1, 2]. The depth of the channels is usually of the same order of magnitude (∼10–100 μm), while channel lengths may be much larger (up to ∼500× the width, that is, 5–50 mm long).

One key benefit of miniaturization is the dramatic reduction in the required fluid sample volume: a linear reduction in the characteristic dimension of the device (L) by a factor of 10^3 (e.g., from 1 cm to 10 μm) amounts to a volume reduction by a factor of 10^9 (L^3). In microfluidic devices, the sample volumes required to fill up a channel typically range from the microliter scale down to the nanoliter scale. Furthermore, as a consequence of miniaturization, high surface-to-volume ratios are observed in microfluidic devices, as illustrated in Figure 6.1.

The high surface-to-volume ratios typical of microfluidics imply that the balance between surface forces (e.g., due to viscous friction and surface tension) and volume forces (e.g., inertia, gravity) is shifted toward the former. This represents a major difference relative to macroscale flows, and is crucial for several practical applications. For example, it is possible to fill up a microchannel by capillarity, which would be unthinkable in a macro device – this principle is commonly used in commercial systems, such as glucose and cholesterol meters to lead the blood droplet through the capillary in the test strip where a chemical reaction takes place.

Both macro- and microfluidic flows are commonly driven by pressure gradients and these are frequently induced using pumps. In microfluidics, special positive displacement pumps, such as syringe pumps, are typically employed to pump the fluid through the device. Alternatively, electro-osmosis (EO) can be used to drive and control liquid flows, provided the fluid contains ions. Electrokinetic flows have been used for a long time in colloidal and porous systems [3, 4], but have only really come of age in microfluidics. The formation of an electric double layer (EDL) allows electrically conductive fluids to be moved in the microchannels by EO (e.g., [5, 6]). The microchannel walls (as most solid surfaces) acquire an electric charge when in contact with an electrolyte (e.g., water) – an EDL of counter-ions will form spontaneously at the walls by attracting nearby counter-ions and repelling co-ions. When an electric potential is applied across the channel, the ions in the EDL move in the direction of the electrode of opposite polarity. This causes a motion of the fluid near the walls, which in turn creates an advective motion of the bulk fluid through viscous forces. The fluid motion exhibits a plug-like profile instead of the characteristic parabolic velocity profile of pressure-driven flows (PDF). Once more, electro-osmotic flows (EOF) are effective at the microscale because of the dominance of surface effects relative to volume effects. In addition to EO, there are other electrokinetic effects important at the microscale, namely electrophoresis, sedimentation potential, and streaming potential. These concepts are thoroughly reviewed by Bruus [6] and there are many other interesting references and reviews available for electrokinetic effects in microfluidic devices (e.g., [7–11]).

The relative balance between inertial and viscous forces is normally quantified in terms of the dimensionless Reynolds number, defined as

$$Re = \frac{\varrho UL}{\eta} \qquad (6.1)$$

where L is a characteristic dimension of the channel, U is a characteristic velocity, usually the average velocity, and ϱ and η are the density and shear viscosity of the fluid, respectively. The magnitude of the Reynolds number is useful to identify the flow regime – laminar or turbulent. The reduced length scales and the dominance of viscous forces over inertial forces means that the flows in microfluidic channels are typically characterized by low to moderate Reynolds numbers (usually smaller than 100, and often smaller than 1). At these low Reynolds numbers, the flow is laminar and no turbulence occurs in contrast to what is usually found at the macroscale. Indeed, for laminar flow to be achieved at the macroscale, highly viscous fluids or very low velocities must be employed, whereas at the microscale, laminar flows can be readily achieved even with low viscosity fluids such as water. This is a major change relative to classical transport processes at the macroscale, and may be an advantage or a disadvantage, depending on the particular application in mind. A number of new technological applications have emerged to take advantage of the laminar behavior of the flow, such as bioassays [12, 13], sorting and separating products of a reaction [1], or microfabrication using UV laminar flow patterning [14]. Conversely, many applications require intense mixing, which can be easily (and rapidly) achieved at the macroscale as fluids mix advectively under high inertia flow conditions, but not so at the microscale where mixing relies mainly on diffusion. Nevertheless, even at Reynolds numbers below 100 it is possible to enhance mixing on the basis of momentum phenomena such as flow separation as well as viscoelastic flow instabilities [15]. The latter will be further discussed in this chapter.

Microfluidic systems have a number of other characteristics that can act as advantages or challenges depending on the application. For instance, a small consumption of reagents can be translated into significant savings both in terms of cost and time. This is critical for many applications, namely in biotechnology, when the samples to be used are costly or available only in limited amounts (e.g., blood), or when a large number of samples are needed, for example, in high-throughput screening [16]. Conversely, in applications that involve the detection of biomolecules, as the volumes are reduced, the detection signals become weaker and consequently new detection methods (and improved labels when appropriate) need to be developed for use at the microscale [17]. Furthermore, as the volume-to-surface ratio decreases, liquid evaporation can become an issue if the processes are slow and occur at high temperatures. Other advantages that arise as a consequence of the reduced length scales include significant waste reduction; reduced cost of fabrication; and possibility of producing highly integrated, disposable, and portable devices. The portability of microfluidic devices results from a combination of the small sizes involved and the low energy consumptions, which makes this technology suitable for wireless solutions [18]. On the other hand, one of the main problems in microfluidics is that the design and fabrication of components are technologically challenging and in most cases cannot simply rely on a scaled down version of their macroscale counterparts [15]. The effort spent in developing efficient microcomponents is well apparent in the number of publications dedicated to development of micropumps, micromixers, and so on (cf. reviews [10, 19, 20] and references therein). Like component design, other difficulties in dealing with microfluidic systems are often a consequence of its youth and can potentially be overcome by further research and development. Figure 6.2 summarizes the main characteristics of microfluidic systems as well as the resulting opportunities and challenges associated with fluidic miniaturization.

The advantages identified, together with recent developments in microfabrication techniques that allow for inexpensive and rapid manufacture of high-quality geometries with well-defined micron-sized features [21–23], have stimulated a remarkable

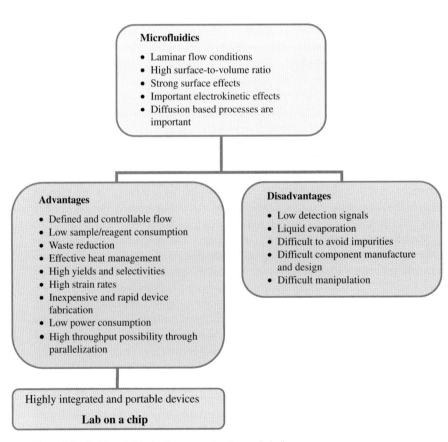

Figure 6.2 Fluidic miniaturization: opportunities and challenges.

growth and found an extensive range of applications in science and technology, as in biology, medicine, and engineering [24]. The printing heads of inkjet printers are one of the most mature commercial applications using microfluidic based systems [25]. Other examples include miniaturized systems for production of suspensions and emulsions [26, 27], immunoassays [13, 28], detection of drugs, flow cytometry [29, 30], dynamic cell separation [31, 32], cell/protein patterning [33], single cell analysis [34], manipulation and analysis of DNA molecules [35–38], and fuel cells [39]. Many other applications have been envisioned and the reader is referred to the literature for further details (e.g., [1, 9, 40, 41]).

The commercial impact of microfluidics is becoming increasingly significant and microfluidic research aspires to have an impact in the automation of biology and chemistry comparable to the microchip in electronics [1, 42]. Considering only applications in the areas of life sciences and *in-vitro* diagnostics, the market value reached 500 million Euros in 2008, and is projected to exceed 2000 million Euros in 2014 [43]. More importantly, it is anticipated that the unique characteristics of microfluidic systems have the potential to trigger a range of novel applications in many areas of science and technology [24]. One of the greatest envisaged microfluidic technological applications consists of a miniaturized laboratory where multiple processes can be integrated into a portable platform known as a lab-on-a-chip. Ultimately, this would correspond to shrinking a full production plant or an analysis laboratory into a small chip [44].

6.1.2.2 Complex Fluids in Microfluidic Flows

Many of the applications mentioned in the previous section involve handling fluids that have a complex microstructure such as polymeric solutions, whole blood or protein solutions. The flow of these fluids may prompt non-Newtonian behavior and in particular viscoelasticity [45, 46]. For instance, fluids with large polymeric molecules often exhibit elastic behavior due to the stretching and coiling of the polymeric chains, which significantly enrich flow behavior [45]. For the characterization of flows with viscoelastic fluids, in addition to the Reynolds number it is important to quantify the Deborah number, De, the Weissenberg number, Wi, and the Elasticity number, El. The Deborah number is defined as the ratio between the relaxation time of the fluid (λ) and the time of observation of the flow (t_f), like the duration of the unsteady part of a flow:

$$De = \lambda/t_f \tag{6.2}$$

The Weissenberg is defined as the product of the relaxation time and a characteristic rate of deformation of the flow (U/L), and quantifies the nonlinear response of the fluid

$$Wi = \lambda U/L \tag{6.3}$$

while El represents the ratio between elastic and inertial effects

$$El = \frac{Wi}{Re} = \frac{\lambda \eta}{\varrho L^2} \tag{6.4}$$

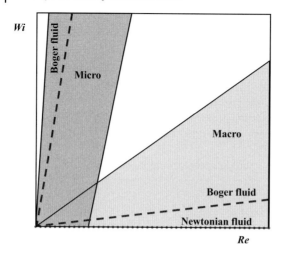

Figure 6.3 Operational regions in the Wi–Re parameter space. The dotted line corresponds to Newtonian fluids (Wi = 0) and the dashed lines represent a Boger fluid (i.e., viscoelastic fluid with constant shear viscosity, cf. Section 6.3) with low viscosity and low relaxation time in flows at the micro- and macroscale.

In steady Eulerian flows with unsteady Lagrangian characteristics, such as the flow in a contraction, the Weissenberg and Deborah numbers are proportional and, as pointed out by Dealy [47], there has been widespread misapplication of both dimensionless numbers. The small length scales together with the high deformation rates and short transit times characteristic of microfluidic systems enable the generation of high Deborah or Weissenberg number flows while keeping the Reynolds number low, leading to high El flows. These distinctive flow conditions result in the ability to promote strong viscoelastic effects, which are not masked by fluid inertia, even in low viscosity/elasticity fluids that would in contrast exhibit Newtonian-like behavior at the equivalent macroscale [48–52]. The dimensionless Wi–Re parameter space is depicted in Figure 6.3, where the operation regions for macro- and microscale flows are distinguished. It is clear that the geometric scale of microfluidic devices results in flows that are distinct from those seen at the macroscale, particularly when they are extension dominated [48, 49, 52–55].

6.1.2.3 Continuum Approximation

We end this introduction by analyzing the validity of the continuum approximation for modeling fluid flow at the microscale. The continuum approximation implies that fluid and flow properties (such as density, viscosity, velocity, stresses, etc.) are defined everywhere in space and vary continuously throughout space [56]. Flows can be modeled by the continuum approximation, also using molecular dynamics, which considers a collection of individual interacting molecules, or more recently as a combination of both approaches using multiscale techniques [57, 58]. Adopting the continuum approach is generally much simpler, it easily considers large systems and is less time consuming than the other techniques, which are still not feasible for many realistic applications and for a sufficiently large number of molecules [42]. However, in simplified terms, for the continuum approximation to hold two main conditions need to be met: (i) the molecules need to be small enough compared to the characteristic length scale of the flow; (ii) the number of molecules inside each fluid element needs to be large enough. In classical fluid mechanics at the macroscale, these conditions are generally satisfied and the continuum approach generally holds [56].

The same is also true in many microfluidics systems, especially those operating with liquids. For example, in Newtonian liquid flows at micrometer-length scales it has been well established that under standard conditions the basic continuum laws governing fluid flow, expressed by the mass conservation and the momentum equations, and the no-slip boundary condition at walls, remain valid [25, 51, 58–60]. For water, the continuum assumption is not expected to break down when the channel dimensions are above 1 μm [5]. For molecules such as water, the ratio of molecular size ($\sim$0.3 nm) to geometric length scale (typically on the order of tens to hundreds of microns) is $\sim 10^{-5}$–10^{-6}. As such, it is considered that there are enough molecules at each location within the flow (the concept of fluid particle as a small volume with a large number of molecules is useful) and that the molecules are small enough to treat the flow under the continuum theory [24]. This remains valid even for more complex fluid flows, including high-molecular-weight polymeric solutions, as attested by the agreement between experimental and numerical data in microfluidics, which provides further credibility to this assumption [55, 61, 62].

However, there are a number of exceptions to the validity of the continuum hypothesis as the characteristic length scales of the flow decrease significantly [63, 64], namely when considering gas flows or gas–liquid flows, in which the gas density is very low compared to liquids. In gas flows, the Knudsen number representing the ratio between the mean free path of molecules and the characteristic length scale of the flow is used to evaluate the validity of the continuum approach. Based on the experimental evidence, it is generally accepted that for Knudsen numbers below 0.01 the continuum approximation is valid. For Knudsen numbers above 0.01, there are deviations to the continuum theory, which are handled initially with corrections and subsequently by other theories that describe microscale flow [57, 58, 65].

The other notable exception is related to complex fluids that are composed of large particles in suspension (e.g., red blood cells) or long molecules such as DNA or even polymers of high molecular weight. The radius of gyration of a polymer chain or the characteristic radius of a suspended particle typically varies from 1 nm to 10 μm. As such, for particle/molecule sizes in the high end of the range, assuming a continuum can be misleading since the working fluid may not be well approximated as microstructurally homogeneous [66]. In this case, other methods should be used to properly model the flow.

Although it is important to be aware of cases where the validity of the continuum approximation breaks down, in all situations of relevance to this chapter, the typical dimensions of molecules and channels are within the range of application of the continuum approach.

6.2
Mixing in Microfluidics

6.2.1
Challenges of Micromixing

Efficient mixing may be defined as a procedure for homogenizing an otherwise inhomogeneous system in the shortest possible amount of time and using the least amount of energy [67]. Mixing is required for many practical applications, in particular in association with chemical reaction. Furthermore, rapid mixing is often an essential requirement to achieve a good performance in many microfluidic applications, namely for biochemistry analysis, drug delivery, sequencing and synthesis of nucleic acids, protein folding, and chemical analysis or synthesis.

In macroscale devices, fluid mixing can often be readily achieved by inducing turbulent flow. In contrast, though not impossible, turbulence is more difficult to reach in microfluidic systems due to the reduced length scale of the channels. Additionally, in many microfluidic applications associated with biological systems, the velocity of the flow cannot be too high since high velocities may lead to large shear stresses that can damage cells and compromise their function [15]. Therefore, in the large majority of cases, microfluidic flows take place in the laminar regime, and often at low Reynolds numbers.

The steady laminar flow of Newtonian fluids in ducts is deterministic. When the Reynolds numbers are low, fluids do not mix advectively when different streams come together in a straight microchannel. Instead, the fluid streams flow in parallel as shown in Figure 6.4, with mixing occurring only due to molecular diffusion across the interface between the streams. At this point, it is useful to introduce the dimensionless Péclet number, which expresses the relative importance of the convective over the diffusive mass transport

$$Pe = UL/D \qquad (6.5)$$

where D is the diffusion coefficient. For typical microfluidic flow conditions, Pe is generally higher than 10, which means that the diffusion process acts more slowly

Figure 6.4 Junction of two Newtonian fluid streams in a microfluidic device under low Re flow conditions.

than the hydrodynamic transport. Additionally, advection is often parallel to the main flow direction and is not useful for the transversal mixing process [19].

Considering a two-dimensional system for simplicity, the mean residence time of a fluid element in the channel, t_R, can be estimated as the ratio between the length of the channel, L, and the average velocity, U,

$$t_R = L/U \tag{6.6}$$

and the time for diffusion (t_D), that is, the time a molecule takes to diffuse a distance d, can be estimated as

$$t_D = d^2/D \tag{6.7}$$

In general, the smaller the molecule, the larger the diffusion coefficient and the faster the molecule can diffuse. Diffusion coefficients for common liquids are quite low (as compared to gases, for example) and can vary widely. For example, small ions in water have diffusion coefficients around $D = 2 \times 10^{-9}$ m^2 s^{-1}, while a large molecule like hemoglobin (in an aqueous solution) has a diffusion coefficient more than two orders of magnitude lower $D = 7 \times 10^{-12}$ m^2 s^{-1}. Thus, small ions take around 5 s to diffuse 100 μm in water, while hemoglobin takes almost 25 min to diffuse over the same 100 μm.

Besides the diffusion coefficient, the other crucial parameter to evaluate the mixing time due to diffusion is the relevant length for mixing (cf. Eq. (6.7)). For example, a protein of 70 kDa requires only 1 s to diffuse 10 μm but more than 10 days to diffuse 1 cm [16]. Taken together, these two effects very often imply that mixing times due to diffusion can be very long relative to the residence time of the fluid in the microchannel. Increasing the channel length implies increasing the pressure drop across the channel and therefore the requirements for micropumping and channel structural strength become more demanding [68]. Additionally, in many reactive systems, having such long mixing times/lengths is not admissible and alternative solutions must be sought.

In summary, liquid mixing at the microscale is not a straightforward task [9] as typical length scales of microfluidic devices are too small to experience mixing induced by turbulence and often too large for diffusion to happen fast enough to provide adequate means of mixing [33, 69]. This means that in most cases, alternative strategies must be implemented for micromixing enhancement.

6.2.2
Overview of Methods for Micromixing Enhancement

Since mixing by molecular diffusion is generally not efficient, other mechanisms need to be brought into action, such as secondary flows due to fluid nonlinearities, flow instabilities, or external actuators. These may be categorized into passive and active methods. Active mixers use external sources to increase the interfacial area between fluid streams, while passive mixers rely on fixed geometrical features (i.e., there are no moving parts) [33], utilize no external energy input, and depend largely on the mechanism used for generating fluid flow through the microchannel [24]. A good introduction to the general theme of mixing is presented by Ottino [70], and by Nguyen [15] for the particular case of micromixing, and is only briefly summarized below.

One possible approach to enhance mixing, inspired by macromixers, is to use active methods to perturb the low Reynolds number flows. Active mixing requires external forcing to induce a flow disturbance and hence increases the amount of transverse flow within the channel. These forces may come from moving mechanical parts and/or external actuators [71]. Active mixers usually produce high levels of mixing, but the systems are considerably more complex, may be difficult to integrate into microfluidic devices, and can be expensive to manufacture [24]. A particular challenge is related to the dominance of surface effects over volume effects as the systems are miniaturized. As a consequence, actuation concepts based on volume forces (e.g., magnetic stirrer), which are widely used at the macroscale, become less efficient at the microscale [15].

The actuator for active mixing can be a pump or work as an energy source, for example in pulsating side flow [72], micropumping for stopping and restarting the flow [73], application of unsteady electric fields acting on the fluid or on suspended particles [74], application of potential differences across pairs of electrodes within the microchannel in the presence of an external magnetic field [75, 76], application of thermal gradients to induce disturbances in the flow using either thermopneumatic actuators (based on the thermal expansion of gases), thermal-expansion actuators (based on the thermal expansion of solids) and bimetallic actuators (based on the difference in thermal coefficient of expansion of two bonded solids) [15], or application of acoustic fields [77, 78]. For further details, the reader is referred to [15, 79]. Active principles can also obviously be used in combination with passive techniques.

Another alternative to reduce mixing times is to induce stirring by chaotic advection [80], with final mixing by diffusion, a process that has also been used in microfluidics [81, 82] and requires a non-negligible Reynolds number since chaotic advection is inherently a nonlinear inertial effect. This is usually accomplished in various ways, depending on the flow Reynolds number, but invariably the flow becomes time-dependent and can also be three-dimensional [19]. If the Reynolds number is low and the fluid is Newtonian, the use of 2D obstacles is usually insufficient to create chaotic advection and enhance mixing. Asymmetric and 3D arrangements of flow perturbations, such as grooves, obstacles, and duct twists become necessary to impart the stretching, reorientation, and randomization mechanisms of distributive mixing [19, 83]. Micromixing in Newtonian fluids by chaotic advection is reviewed in detail by Nguyen [15].

Fluids in microsystems very often contain additives that impart non-Newtonian characteristics to the fluids and, in particular, viscoelasticity. These rheological characteristics introduce nonlinearities that can be explored to dramatically change the flow dynamics, and in particular to enhance mixing [1, 54, 84]. The elasticity of the fluids is characterized, among other things, by the appearance of anisotropic normal stresses, which produce secondary flows [85] and/or elastic instabilities even at extremely low Reynolds number. Although weak, these secondary flows help the appearance of flow instabilities and reduce mixing times, because they create conditions similar to those of chaotic advection, that is, 3D flow which we call here chaotic elastic flow (inertia is negligible). The elastic instabilities have been shown to exist even in the absence of inertia and are associated with strong curvature of streamlines and large normal stresses [86]. When the elastic instabilities become very intense, reaching a saturated nonlinear state, fluctuations even become random over a wide range of length and time scales [87], very much like inertial turbulence, in spite of negligible Reynolds numbers. This has prompted Groisman and Steinberg [88] to call it "elastic turbulence." So, elastic effects are used to reduce the critical conditions for the existence of chaotic flow and enhanced mixing, allowing the use, at lower Reynolds numbers, of passive techniques usually associated with higher Reynolds number flows. This type of passive mixing is discussed in detail in Section 6.5. Before that, however, we introduce in Section 6.3 some basic concepts about non-Newtonian fluids, as well as the governing equations required for flows of complex fluids (Section 6.4).

6.3
Non-Newtonian Viscoelastic Fluids

In this section, we present a brief overview of the rheology of non-Newtonian fluids. More detailed descriptions are found in [45, 46], among others. Rheometry is also described in [89, 90].

The rheology of fluids is assessed through their behavior in a small set of controllable (and quasi-controllable) flows, whose kinematics are known and independent of fluid properties. For shear-based properties, this is the Couette flow schematically shown in Figure 6.5 in the planar (2D) version. Technologically, the Couette flow is usually implemented in an axisymmetric version, as in the concentric cylinders, cone–plate, or plate–plate geometries for which the applied torque and rotational speed are directly proportional to the shear stress and shear rate, respectively. The use of small gaps in these geometries ensures a controllable flow and a nearly constant shear rate across the gap. For extensional-based properties the ideal flow is a purely extensional flow, such as the uniaxial extension, but it is not always possible to implement it easily, especially for low-viscosity fluids.

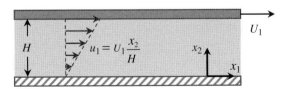

Figure 6.5 Plane Couette flow and coordinate system.

6.3.1
Shear Viscosity

Shear viscosity is defined as the ratio between shear stress (τ_{12}) and the shear rate ($\dot{\gamma}$) in the Couette flow of Figure 6.5, where subscripts 1 and 2 denote streamwise and transverse directions, respectively:

$$\eta = \frac{\tau_{12}}{du_1/dx_2} = \frac{\tau_{12}}{U_1/H} = \frac{\tau_{12}}{\dot{\gamma}} \tag{6.8}$$

Typically, non-Newtonian fluids have a shear-thinning behavior with a low shear rate constant viscosity plateau, as shown in Figure 6.6. A second lower constant viscosity plateau at high shear rates is also frequent, but often this is not observed in rheometric flows before the onset of flow instabilities. Some suspensions of irregular solids, or surfactant solutions, exhibit a shear-thickening behavior, but this is often limited to a narrow range of shear rates.

There are materials for which the first Newtonian plateau of the shear viscosity is not observed, and the shear viscosity grows to infinity at vanishingly small shear rates. These materials possess some form of internal structure for which a minimum stress is required prior to yielding – the yield stress – and often their viscosity depends not only on the shear rate but also on time – thixotropy or anti-thixotropy, depending on whether the shear viscosity decreases or increases over time. Examples are toothpaste, mayonnaise, blood, and suspensions of particles, in which the effect is enhanced if macromolecules are present. Dilute and semidilute polymer solutions do not exhibit yield stress and thixotropy, so these properties will not be considered

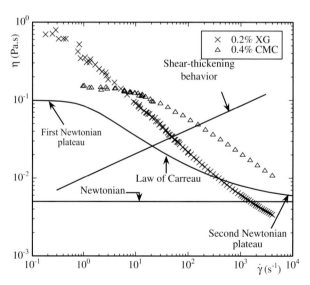

Figure 6.6 Shear viscosity of aqueous solutions of 0.2% by weight xanthan gum (XG) and 0.4% by weight carboxymethyl cellulose (CMC) and typical behavior of some rheological models.

further. The interested reader is referred to Larson [46] and additional papers on issues and techniques involving yield stress fluids [91–95].

6.3.2
Normal Stresses

Viscoelastic fluids develop normal stresses in shear flow, which are known within a constant value, so their differences are the useful material properties. For a pure shear flow as illustrated in Figure 6.5, the first normal stress difference (N_1) is defined as the difference between the streamwise normal stress (τ_{11}) and the transverse normal stress (τ_{22}), and gives rise to the material property designated as first normal stress difference coefficient, Ψ_1:

$$\Psi_1 \equiv \frac{N_1}{\dot{\gamma}^2} = \frac{\tau_{11}-\tau_{22}}{\dot{\gamma}^2} \tag{6.9}$$

The second normal stress difference is $N_2 \equiv \tau_{22}-\tau_{33}$ and the corresponding coefficient is $\Psi_2 = N_2/\dot{\gamma}^2$. N_2 is usually small, with maximum values not exceeding 20% of N_1 and with an opposite sign to N_1. Measurement of N_2 is difficult and can be done using a special cone–plate apparatus [96].

The typical behavior of a viscoelastic fluid regarding Ψ_1 is included in Figure 6.7, which pertains to an aqueous solution of polyacrylamide (PAA) at a weight concentration of 300 ppm [97]. In the limit of small shear rates, Ψ_1 tends to a constant value, to which corresponds $N_1 \rightarrow 0$. So, even though the behavior of Ψ_1 depicted in Figure 6.7 is shear-thinning, the normal stresses grow quickly as N_1 varies with the square of the shear rate (when Ψ_1 is constant). N_1 is responsible for some spectacular phenomena, such as the Weissenberg effect [45]. Today, the capability of measurement of N_1 is standard in commercial rotational rheometers.

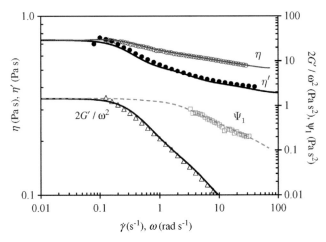

Figure 6.7 Material functions of a 300-ppm PAA solution under steady shear and SAOS flows. Details of fluid composition can be found in [97].

6.3.3
Storage and Loss Moduli

In small amplitude oscillatory shear flow (SAOS) of a viscoelastic fluid, an oscillating strain, $\gamma = \gamma_0 \sin(\omega t)$, is applied to one of the walls of the Couette cell. The resulting shear stress will be given by $\tau(t) = \tau_0 \sin(\omega t + \delta)$, and is out of phase by δ relative to the imposed strain. Provided the amplitude of deformation is small, the response of the material depends only on the forcing frequency and the resulting storage (G') and loss (G'') moduli are mathematically defined as

$$G' = \omega \eta'' \equiv \frac{\tau_0}{\gamma_0} \cos \delta; \qquad G'' = \omega \eta' \equiv \frac{\tau_0}{\gamma_0} \sin \delta \qquad (6.10)$$

which measure the amount of energy stored reversibly by the material (G', deformation in phase with the stress) and consequently can be recovered, and the energy irreversibly lost by viscous dissipation (G'', stress out of phase with the deformation). Sometimes the components η' and η'' of the complex dynamic viscosity (η^*) are used instead, where $\eta^* = \eta' - i\eta''$, with i representing the imaginary number ($i^2 = -1$).

For a Newtonian fluid, the response in this test would be obvious ($G' = 0$, $G'' = \tau_0/\gamma_0$) so the loss angle (δ) would be maximum and given by $\delta = \pi/2$ (note that $\tan \delta = G''/G'$).

6.3.4
Extensional Viscosity

In a pure extensional flow, the velocity vector only varies in its direction, as in a traction or compression experiment. If a fluid sample is subject to an extensional flow, such as the flow in a contraction or in a pulling device (cf. Figure 6.8), it undergoes an extensional deformation and develops normal stresses proportional to the normal strain rate ($\dot{\varepsilon}$). The ratio between the normal stress difference and the strain rate defines the extensional viscosity

$$\eta_E \equiv \frac{\tau_{11} - \tau_{22}}{\partial u_1/\partial x_1} = \frac{\tau_{11} - \tau_{22}}{\dot{\varepsilon}} \qquad (6.11)$$

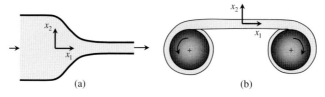

Figure 6.8 Schematic representation of flows with a strong extensional deformation: (a) smooth contraction flow; (b) extensional flow device.

Note that all fluids, including Newtonian fluids, have a nonzero extensional viscosity. For Newtonian fluids, the uniaxial extensional viscosity equals three times the shear-viscosity, so no distinction is required, but for viscoelastic fluids the ratio between the extensional and shear viscosities, called the Trouton ratio, varies with the rate of deformation and can largely exceed the value of three, attaining sometimes values of the order of 100 or higher. An impressive consequence of a very high extensional viscosity is the tubeless siphon experiment [45].

The measurement of the extensional viscosity is not easy, because it is difficult to ensure that fluid particles are under a constant strain rate for a sufficiently long time to eliminate transient start-up effects, especially at high strain rates. Additionally, for the mobile systems of interest here it is difficult to impose a constant strain rate flow and so the extensional viscosity can only be directly measured with such devices as the capillary breakup extensional rheometer (CaBER) [98]. A variant of the CaBER is the filament stretching extensional rheometer (FiSER) based on the work of Tirtaatmadja and Sridhar [99], where the fluid filament between plates is deformed as the plates move with a velocity increasing exponentially with time. This allows the measurement of strain-dependent extensional viscosity [100].

Alternatively, there are flows with a strong extensional nature from which an extensional viscosity indexer can be obtained, such as the pressure drop enhancement in a contraction flow or the tensile force required to sustain fluid stretching in the space between two nozzles in the opposed jet rheometer, but in these flows the fluids are not subject to a constant strain rate and the flow is contaminated by secondary effects that may overwhelm the main measurement. In contrast, the high consistency of polymer melts facilitates the integrity of fluid samples under uniaxial extension and a number of devices can be used to measure their extensional viscosity, such as the Sentmanat device [101].

6.3.5
Other Rheological Properties

The rheological properties discussed can today be reliably measured and are standard. However, it is clear to rheologists and fluid dynamicists alike that the set does not guarantee that if a rheological constitutive equation is able to predict all of them for a particular fluid, it will be able to predict accurately all types of flow with that fluid [102, 103], a situation quite similar to the prediction of Newtonian turbulent flows with turbulence models. This indicates the need for other fluid properties, especially those related to time-dependency and nonlinear effects.

Other tests, such as the creep and the stress relaxation flows in shear and strain, are good examples. One may also consider the response of fluids to a sequence of steps in normal or shear strain, since here the response of fluids is different from that to a single step. To assess nonlinear viscoelasticity, meaningful interpretation of data from large amplitude oscillatory shear flow (LAOS) is currently under development [104].

6.4
Governing Equations

Viscoelastic fluid flow is governed by the momentum and continuity equations together with a rheological constitutive equation adequate for the fluid. If heat transfer is involved, the energy equation must be included with the corresponding thermal constitutive equation, usually Fourier's heat law. To consider chemical reaction, the mass conservation equation for each chemical species needs to be solved in combination with the mass transport constitutive equation, usually Fick's law. To assess mixing performance, it may be necessary to solve a transport equation for an adequate scalar. These equations are coupled in a variety of ways: dependence of fluid properties on temperature, molecular orientation and/or fluid composition, through new terms in the governing equation, such as buoyancy in the momentum equation or extra terms in the constitutive equation, which can be traced back to the effect of temperature on the mechanisms acting at microscopic level. The treatment of these extra terms of the constitutive equations is an advanced topic not considered here. For more in-depth discussions, the reader is referred to [105–107].

In general, the fluid dynamics and heat transfer problems are coupled and the set of governing equations has to be solved simultaneously. For a general flow problem, this can only be done numerically, but under simplified conditions, such as temperature-independent fluid properties (a good approximation, if temperature variations are small), it is possible to solve for the flow without consideration for the thermal problem (although not the other way around). Other times, the solution can still be obtained assuming temperature-independent properties, but a correction is introduced to compensate for the neglected effect. This is a fairly successful approach for simple geometries and simple fluids (such as inelastic fluids), but for viscoelastic fluids a more exact approach may be required for accurate results [108].

The governing equations are presented in the next sections in tensor notation for generality. The reader is referred to the appendices of Bird et al. [45, 109, 110] for an extensive presentation of their form in various coordinate systems.

6.4.1
Continuity and Momentum Equations

The continuity equation is written as

$$\frac{\partial \varrho}{\partial t} + \nabla \cdot (\varrho \mathbf{u}) = 0 \quad (6.12)$$

and the momentum equation as [58]

$$\frac{\partial (\varrho \mathbf{u})}{\partial t} + \varrho (\mathbf{u} \cdot \nabla) \mathbf{u} = -\nabla p + \varrho \mathbf{g} + \nabla \cdot \tau_t + \varrho_e \mathbf{E} - \frac{1}{2} \mathbf{E} \cdot \mathbf{E} \epsilon_0 \nabla \epsilon + \frac{\epsilon_0}{2} \nabla \left(\varrho \frac{\partial \epsilon}{\partial \varrho} \mathbf{E} \cdot \mathbf{E} \right) \quad (6.13)$$

where $\mathbf{u}$ is the velocity vector, p is the pressure, ϱ is the fluid density, and the fluid total extra stress (τ_t) is given by an adequate rheological constitutive equation. The last three terms on the right-hand side take electrokinetic effects into account, where ϱ_e denotes the net electric charge distribution within the fluid, $\mathbf{E}$ represents the applied electric field (or induced streaming potential in flows with electroviscous effects), ϵ_0 is the dielectric permittivity of vacuum, and ϵ is the dielectric constant of the fluid. The last term accounts for permittivity variations with fluid density and is only relevant at gas–liquid interfaces or in ionized gas flows, whereas the penultimate term accounts for spatial variations in the dielectric constant of the fluid. Thus, for incompressible fluids of constant dielectric permittivity only the first of the three terms is required, which is known as Lorentz force.

In this chapter we will assume that the imposed electrical potential and the induced potential are independent of each other and therefore they can be linearly combined into the total electric potential $\Phi = \phi + \psi$, so that the resulting electrical field is $\mathbf{E} = -\nabla\Phi$. This is admissible when the EDL is thin, and also requires a weak applied streamwise gradient of electrical potential, that is, $\Delta\phi/L \ll \psi_0/\zeta$, where $\Delta\phi$ is the potential difference of the applied electrical field, L is the distance between the electrodes, and ζ is the Debye layer thickness. In this case, the transverse charge distribution is essentially determined by the potential at the wall, ψ_0, the so-called zeta potential. If the local EOF velocities are small and/or parallel to the walls, as in thin EDLs, the effect of fluid motion on the charge distribution can also be neglected. These simplifications are part of the so-called standard electrokinetic model, in which case Eq. (6.13) becomes

$$\frac{\partial(\varrho \mathbf{u})}{\partial t} + \varrho(\mathbf{u} \cdot \nabla)\mathbf{u} = -\nabla p + \varrho \mathbf{g} + \nabla \cdot \tau_t - \varrho_e \nabla \Phi \tag{6.14}$$

6.4.2
Rheological Constitutive Equation

The fluid total extra stress (τ_t) is given as the sum of an incompressible solvent contribution having a viscosity coefficient η_s and a polymer/additive stress contribution τ_p, as

$$\tau_t = 2\eta_s(II_\mathbf{D}, III_\mathbf{D})\mathbf{D} + \tau_p \tag{6.15}$$

The solvent viscosity coefficient in Eq. (6.15) has been made to depend on the second and third invariants ($II_\mathbf{D}$, $III_\mathbf{D}$) of the rate of deformation tensor $\mathbf{D}$ defined as

$$\mathbf{D} = \frac{1}{2}\left(\nabla \mathbf{u} + \nabla \mathbf{u}^T\right) \tag{6.16}$$

to consider both the possibility of having a Newtonian (constant viscosity) or a non-Newtonian (variable viscosity) solvent. In this way, Eq. (6.15) includes the class of inelastic non-Newtonian fluids known as generalized Newtonian fluids (GNF) for which the polymer contribution is set to zero ($\tau_p = 0$). Then, the viscosity coefficient depends on invariants of the rate of deformation tensor, the most common being the second invariant, defined in the next section. For viscoelastic fluids, η_s is set to zero for polymer melts or to a nonzero constant when dealing with a polymer solution based on a Newtonian solvent.

Usually the non-Newtonian fluids are treated as incompressible fluids, so the continuity equation simplifies to $\nabla \cdot \mathbf{u} = 0$. Some very limited phenomena may require consideration of liquid compressibility, an issue not considered here.

6.4.2.1 Generalized Newtonian Fluid Model

The purely viscous GNF model is defined in Eq. (6.15) with $\tau_p = \mathbf{0}$ and the fluid viscosity $\eta_s(II_\mathbf{D}, III_\mathbf{D})$ depending on invariants of the rate of deformation tensor [111]. The most common viscosity models consider only dependence on the second invariant and we can write many of them in a compact form as

$$\eta_s(II_\mathbf{D}) = (\eta_1 - \eta_2)[\alpha + (\Lambda II_\mathbf{D})^a]^{\frac{n-1}{a}} + \eta_2 \quad \text{with} \quad II_\mathbf{D} \equiv \sqrt{2\mathbf{D}:\mathbf{D}} \tag{6.17}$$

Equation 6.17 includes the Newtonian fluid model (viscosity coefficient η), the Ostwald-de Waele power law (consistency index K and flow behavior index n), the Carreau–Yasuda model (zero shear viscosity η_0, infinite shear rate viscosity η_∞, power index n, transition coefficient a, and transition time scale Λ), the simplified Carreau model and the Sisko model, with the corresponding coefficients given in Table 6.1.

6.4.2.2 Viscoelastic Stress Models

The previous constitutive models cannot predict viscoelastic characteristics, such as any shear-induced normal stresses in Couette flow, or memory effects. There is a class of models, which is still explicit on the stress tensor that can predict some of these elastic effects. One such model, the Criminale– Ericksen– Filbey (CEF) equation, should only be used in steady shear flow in which case it provides accurate results [45]. The CEF model can be written as

$$\tau = 2\eta(\dot{\gamma})\mathbf{D} - \Psi_1(\dot{\gamma})\stackrel{\nabla}{\mathbf{D}} + 4\Psi_2(\dot{\gamma})\mathbf{D}^2 \tag{6.18}$$

with $\dot{\gamma} \equiv II_\mathbf{D}$ and $\stackrel{\nabla}{\mathbf{D}}$ representing the upper-convected derivative of $\mathbf{D}$, defined as

$$\stackrel{\nabla}{\mathbf{D}} \equiv \frac{\partial \mathbf{D}}{\partial t} + (\mathbf{u} \cdot \nabla)\mathbf{D} - \nabla \mathbf{u}^T \cdot \mathbf{D} - \mathbf{D} \cdot \nabla \mathbf{u} \tag{6.19}$$

Table 6.1 Values of parameters in generalized viscosity function of Eq. (6.17) for some typical viscosity models.

	n	a	α	η_1	η_2	Λ	*
Newtonian	1	Any	Any	η	Any	Any	
Power law	n	Any	0	*	0	*	$K = \eta_1 \Lambda^{n-1}$
Carreau–Yasuda	n	Any	1	η_0	η_∞	Λ	
Simplified Carreau	n	2	1	η_0	0	Λ	
Sisko	n	Any	0	*	η_∞	*	$K = (\eta_1 - \eta_2)\Lambda^{n-1}$

Other stress explicit models for viscoelastic fluids are contained in Eq. (6.18), such as the second-order fluid (constant η, Ψ_1, and Ψ_2) or the Reiner–Rivlin equation ($\Psi_2 = 0$). The use of these models should be restricted to weakly elastic fluids and low Weissenberg number flows, that is, to fluids deviating slightly from Newtonian and to slow flows, since outside these conditions they lead to physically incorrect predictions. So, these models are essentially useful to investigate deviations from the behavior of Stokes flows.

More useful are the integro-differential viscoelastic fluid models. The polymeric contribution to the extra-stress tensor in Eq. (6.15) can in general be represented as a set of N modes

$$\tau_p = \sum_{k=1}^{N} \tau_k \qquad (6.20)$$

where each polymer mode obeys a rheological equation of state of integral or differential nature. An example of the latter is the following general equation:

$$f(\text{tr } \tau)\tau + \frac{\lambda}{F(\text{tr } \tau, L^2)} \overset{\triangledown}{\tau} + \frac{\alpha\lambda}{\eta_p}\tau^2 = 2\eta_p \mathbf{D} \qquad (6.21)$$

which includes such models as the upper-convected Maxwell (UCM) model, the Phan-Thien–Tanner model (PTT), the Johnson–Segalman (JS) model, the Giesekus model or the FENE-MCR model, according to Table 6.2. For conciseness and since very often a single mode is used, the subscript indicating the mode has been dropped. Note that for each mode the model parameters can have different numerical values.

Function $f(\text{tr } \tau)$ takes either the exponential form, $f(\text{tr } \tau) = \exp[(\varepsilon\lambda/\eta_p)\text{tr } \tau]$, or a simpler linearized form $f(\text{tr } \tau) = 1 + \left(\varepsilon\lambda/\eta_p\right)\text{tr } \tau$. The temperature influences exponentially the polymer viscosity coefficient, η_p, and the relaxation time, $\lambda = \lambda(T_0)a_T$ where T_0 is a reference temperature, and a_T is the nondimensional shift factor, usually described using the Williams–Landel–Ferry (WLF) equation [112]. The shear modulus, $G = \eta_p/\lambda$, is only weakly dependent on the temperature, as discussed by Wapperom et al. [113]. The same correction for temperature is valid for the material functions in the constitutive equation (6.17) and (6.18). $F(\text{tr } \tau, L^2)$ is the

Table 6.2 Model parameters of Eq. (6.21) for some viscoelastic constitutive equations.

Models	ε	α	L^2	ξ	β
UCM	0	0	∞	0	0
Oldroyd-B	0	0	∞	0	]0, 1[
PTT[a]	>0	0	∞	[0, 2]	0[b]
FENE-MCR	0	0	$\gg 3$	0	]0, 1[
Giesekus	0	]0, 1[	∞	0	0[b]

a) If $\xi = 0$ it is also called the simplified PTT (sPTT) model. The original PTT relies on the exponential form of $f(\text{tr}\tau_p)$, a linearized form uses the linear version of $f(\text{tr}\tau_p)$.
b) Strictly speaking $\beta = 0$ for the PTT or Giesekus models. Today their use is widespread to model polymer solutions with a solvent contribution ($\beta \neq 0$) and the designation stands.

stretch function that depends on the trace of the stress tensor and on the extensibility parameter L^2, representing the ratio of the maximum to the equilibrium average dumbbell extensions for a FENE-MCR model (from finitely extensible nonlinear elastic, with the Chilcott–Rallison approximation) [114]. The stress coefficient function $f(\text{tr }\tau)$ introduces the dimensionless parameter ε, which is closely related to the steady-state elongational viscosity in extensional flow ($\eta_E \propto 1/\varepsilon$ for low ε), while α is the dimensionless mobility factor of the Giesekus equation. Finally, $\overset{\square}{\tau}_p$ denotes the Gordon–Schowalter derivative of the extra-stress tensor, which is a mixture of the upper ($\xi = 0$) and lower ($\xi = 2$) convected derivatives, and is defined as

$$\overset{\square}{\tau} = \frac{D\tau}{Dt} - \tau \cdot \nabla \mathbf{u} - \nabla \mathbf{u}^T \cdot \tau + \xi(\mathbf{D} \cdot \tau + \tau \cdot \mathbf{D}^T) \tag{6.22}$$

Parameter ξ accounts for the slip between the molecular network and the continuum medium and provides nonzero second normal stress differences in pure shear flow. However, the use of $\xi \neq 0$ can lead to unphysical behavior of the model, which are called Hadamard instabilities, if the solvent contribution is weak or nonexistent. β in Table 6.2 denotes the solvent ratio, defined as $\beta = \eta_s/(\eta_s + \eta_p)$.

The UCM model is the simplest viscoelastic differential model and is characterized by a constant shear viscosity, equal to η_p, a constant first normal stress difference coefficient ($\Psi_1 = 2\eta_p \lambda$), and a zero second normal stress difference ($N_2 = 0$). Note that the UCM model requires the solvent viscosity in Eq. (6.15) to be set to zero ($\beta, \eta_s = 0$). If the solvent viscosity is a nonzero constant ($\eta_s \neq 0$), we have the so-called Oldroyd-B model, which has the same elastic properties as the UCM model, whereas for the viscous properties it suffices to add the contribution from the Newtonian solvent. The normal stresses/extensional viscosities of the UCM and Oldroyd-B fluid become unbounded in extensional flow when the rate of deformation tends to $1/(2\lambda)$ as is clear from the steady-state uniaxial extensional viscosity given by

$$\eta_E = 3\eta_p \frac{1}{(1+\lambda\dot{\varepsilon})(1-2\lambda\dot{\varepsilon})} + 3\eta_s \tag{6.23}$$

Nevertheless, these two models contain many of the essential features of viscoelasticity and for this reason they are still extensively used, especially in the development of numerical methods or in preliminary calculations with viscoelastic fluids (a robust method for the UCM and Oldroyd-B models is likely to be robust for other constitutive equations). Additionally, the Oldroyd-B model is adequate to describe the behavior of Boger fluids (constant viscosity elastic fluids). These are mostly dilute polymer solutions in high-viscosity Newtonian solvents, but it is also possible to manufacture them with solvents of moderate viscosity provided these are poor solvents [115].

Regarding the response to SAOS flow, the described viscoelastic models behave identically with their loss and storage moduli given by

$$G' = \eta''\omega = \frac{\eta_p \lambda \omega^2}{1+(\lambda\omega)^2}; \qquad G'' = \eta'\omega = \eta_s \omega + \frac{\eta_p \omega}{1+(\lambda\omega)^2} \tag{6.24}$$

Figure 6.7 shows G' and G'' (via η') as a function of the frequency of oscillation for a 300-ppm aqueous solution of PAA and the corresponding fit by a three-mode polymer model with a Newtonian solvent contribution.

The prediction of variable viscosity and normal stress difference coefficients is provided by the more complex models, such as the PTT, Giesekus, or others. The nonlinear fluid properties are precisely introduced by the nonlinear terms of the equations, with different parameters having different impacts onto the model. Usually, the addition of shear-thinning to the shear viscosity also leads to shear-thinning of Ψ_1 and for $\Psi_2 \neq 0$ it is necessary for the coefficient ξ inside the Gordon–Schowalter derivative to be nonzero, or instead to have the quadratic stress term switched on, as in the Giesekus model.

There are more models for polymer solutions and lately they have been derived on the basis of molecular kinetic theories for polymer molecules, such as the FENE-P model (finitely extensible nonlinear elastic with Peterlin's approximation). For polymer melts, there is also a large set of complex network-based models. All modern constitutive equations have a complex formulation, frequently introducing the concepts of conformation tensor, or of stretch and orientation tensors, among others. As an example, we give below the constitutive equation for the FENE-P model written in terms of the conformation tensor $\mathbf{A}$, which up to a scaling factor corresponds to the second moment of the distribution function of the end-to-end vector of the model dumbbell, $< \mathbf{QQ} >$, via [107]:

$$\tau_p = \frac{\eta_p}{\lambda}[F(\text{tr } \mathbf{A})\mathbf{A} - a\mathbf{I}] \tag{6.25}$$

with

$$F(\text{tr } \mathbf{A})\mathbf{A} + \lambda \stackrel{\nabla}{\mathbf{A}} = a\mathbf{I} \quad \text{and} \quad F(\text{tr } \mathbf{A}) = \frac{L^2}{L^2 - \text{tr } \mathbf{A}} \tag{6.26}$$

where L^2 represents the maximum extensibility of the dumbbell, and $a = 1/(1 - 3/L^2)$.

For more details and models, see the works of Larson [116], Bird et al. [45, 109], and more recently Huilgol and Phan-Thien [117], Larson [46], and Tanner [103].

6.4.3
Equations for Electro-Osmosis

To solve Eq. (6.14) for electrically driven flows, it is necessary to determine the electric charge distribution density. Figure 6.9 illustrates the principle of EO in a simple channel. Basically, when a polar fluid is brought in contact with a surface chemical equilibrium leads to a spontaneous charge being acquired by the wall and simultaneously by the layers of fluid nearer to the surface (with ions of opposite sign, the counter-ions), thus forcing the formation of a near-wall layer of immobile ions followed by a second layer of mobile ions, both of which contain a higher concentration of counter-ions as the co-ions are repelled by the wall [118]. The layer of immobile ions, the Stern layer, and the immediate layer with mobile ions, the diffuse layer, form together the so-called EDL. EOF is obtained when an external electric field

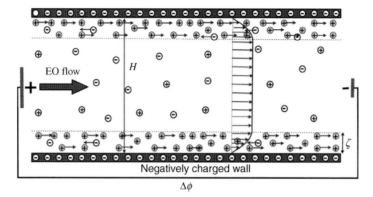

Figure 6.9 Illustration of EO driven flow. The blue and red arrows are Coulombic repulsive and attractive forces on the counter and co-ions, respectively. Adapted from [118].

$\mathbf{E} = -\nabla \phi$ (ϕ is the potential in the streamwise direction) is applied between the channel inlet and outlet thus creating Coulomb forces acting on the charges within the EDL. The motion of these ions drags the remaining fluid laying outside the EDL along the channel. To determine the Coulomb force (last term on the right-hand side of Eq. (6.14)), it is necessary to quantify the net electric charge density, ϱ_e, which is given by

$$\varrho_e = e \sum_i z_i n_i \tag{6.27}$$

where e is the elementary charge, n_i is the bulk number concentration of positive/negative ion i, and z_i is the corresponding ion valence. Note that the bulk number ionic concentration n is related to the molar concentration of ions (c_i) in the electrolyte solution via $n_i = N_A c_i$, where N_A is Avogadro's number [4]. The simplest case is that of electrolytes with equally charged ions of valence $z^- - z^+$ for which the above general Eq. (6.27) simplifies to $\varrho_e = e z(n^+ - n^-)$.

The spontaneously induced potential ψ near the interface/wall is given by

$$\nabla^2 \psi = -\frac{\varrho_e}{\epsilon} \tag{6.28}$$

whereas the imposed streamwise potential is such that

$$\nabla^2 \phi = 0 \tag{6.29}$$

To determine the ionic concentration, their transport equations, also called the Nernst–Planck equations, need to be solved. These are expressed as [6]

$$\frac{\partial(n^\pm)}{\partial t} + \mathbf{u} \cdot \nabla n^\pm = \nabla \cdot \left(D^\pm \nabla n^\pm \right) \pm \nabla \cdot \left[D^\pm n^\pm \frac{ez}{k_B T} \nabla(\phi + \psi) \right] \tag{6.30}$$

where $D^\pm$ are the diffusion coefficients of the $n^\pm$ ions, respectively, k_B is Boltzmann's constant, and T is the absolute temperature. Simpler models can be used in simpler situations: when the flow is essentially unidirectional, steady, and parallel to walls, the ionic distribution becomes stationary and the EDL is restricted to the wall vicinity, so significant variations of $n^{\pm}$ and ψ only occur in the direction normal to the wall and in its vicinity. Then, the Nernst–Planck equations reduce to the stable Boltzmann distribution and the corresponding electric charge density is given by

$$\varrho_e = -2\,n\,e\,z\sinh\left(\frac{ez}{k_BT}\psi\right) \tag{6.31}$$

Equations (6.28) and (6.31) constitute the so-called Poisson–Boltzmann model, which is still quite general. When the ratio between the electric to thermal energies is small, synonymous of a small value of $ez\psi_0/(k_BT)$ (ψ_0 is the zeta potential), the hyperbolic sine function can be linearized ($\sinh x \approx x$) and the electric charge density becomes

$$\varrho_e = -\,\epsilon\,\kappa^2\psi \tag{6.32}$$

where $\kappa^2 = 2e^2z^2n/(\epsilon\,k_BT)$ is the Debye–Hückel parameter related to the thickness of the EDL, $\zeta = 1/\kappa$. Equations (6.28) and (6.32) constitute the Poisson–Boltzmann–Debye–Hückel model.

6.4.4
Thermal Energy Equation

For nonisothermal flows, it is necessary to include in the set of governing equations the following special form of the energy equation:

$$\varrho c\frac{DT}{Dt} = -\nabla\cdot\mathbf{q} + \dot{q}_1 + \tau_t : \mathbf{D} \tag{6.33}$$

where c is the specific heat of the fluid, $\mathbf{q}$ is the conduction heat flux to be quantified below, and $\dot{q}_1$ is a source, here representing Joule heating per unit volume. The last term on the right-hand side represents the mechanical energy supply by the viscoelastic medium (the viscoelastic stress work), which includes the viscous dissipation. This is an important term since many non-Newtonian viscoelastic fluids are highly viscous and have non-negligible internal viscous dissipation, which precludes an isothermal approach. The small channel dimensions in microfluidics, if coupled with large fluid velocities, lead to large shear rates, and the viscoelastic stress work becomes non-negligible.

In rigorous terms, the last term of Eq. (6.33) should have been multiplied by a coefficient κ and an extra term multiplied by $(1-\kappa)$ should have been added to the energy equation in order to account for internal energy storage by the viscoelastic medium [107]. The connection between viscoelasticity and thermal energy and the more specific issue of the numerical value of κ are still topics of research [119] and numerical simulations of Peters and Baaijens [120] have also shown that the results from such an extended equation for viscoelastic fluids are not too different from those obtained with the simpler Eq. (6.33), which neglects the extra internal energy storage term (for pure shear flow, the results are actually exactly the same).

For the diffusive heat flux, Fourier's law of heat conduction is assumed with an isotropic thermal conductivity k

$$\mathbf{q} = -k\nabla T \qquad (6.34)$$

For materials possessing some form of orientational order, such as liquid crystals, the thermal conductivity can have an anisotropic behavior and is now a second-order tensor (**k**), in which case the heat flux is given by $\mathbf{q} = -\mathbf{k} \cdot \nabla T$.

The Joule heating effect is a consequence of the application of an electric field across a conductive fluid (as in EO) and is given in complete form by

$$\dot{q}_1 = \frac{1}{\sigma}(\varrho_e \mathbf{u} + \sigma \mathbf{E}) \cdot (\varrho_e \mathbf{u} + \sigma \mathbf{E}) \qquad (6.35)$$

where σ represents the electrical conductivity of the fluid. Under the conditions of validity of the Debye–Hückel approximation in EO, this Joule heating effect is essentially that due to the electric field, because of the very low velocities, so Eq. (6.35) reduces to $\dot{q}_1 = \sigma \mathbf{E} \cdot \mathbf{E}$.

In principle, all fluid properties may depend on temperature and this strongly couples the rheological equation of state and the momentum equation on one side, with the thermal energy equation on the other. There are obvious advantages in considering fluid properties independent of temperature, because the fluid dynamics becomes independent of the thermal energy, simplifying the problem. The thermal energy equation, however, is always coupled with the flow via the velocity field and its gradients; therefore it can never be dealt with independently from the momentum equation.

6.5
Passive Mixing for Viscoelastic Fluids: Purely Elastic Flow Instabilities

6.5.1
General Considerations

As discussed in Section 6.1, the small length scales of microfluidics increase significantly the role of fluid elasticity beyond what can be achieved at the macroscale, and major differences in behavior are expected [1]. Indeed, complex flows of complex fluids often generate flow instabilities, even under inertialess (or creeping) flow conditions (i.e., when $Re \ll 1$), which are typically encountered at the microscale. These are called purely elastic flow instabilities and can play an important role in the context of mixing improvement at the microscale in viscoelastic fluid flows. In this section, we present an overview of elastic flow instabilities and focus on practical examples related to their development and enhancement at the microscale. As discussed in Section 6.1, flows at the microscale can be driven mainly by imposed pressure gradients, which are considered in this section, or using electrokinetic effects, which are considered in Section 6.6.

6.5.2
The Underlying Physics

The remarkable properties of complex fluids arise from the interaction between their molecular structure and the flow. The flow conditions induce a local molecular rearrangement, with the polymer chains being stretched and oriented. This nonequilibrium configuration generates large anisotropic normal stresses, which themselves influence the flow field. This feedback mechanism can lead to flow destabilization, and is more pronounced above the so-called coil–stretch transition that occurs when the strain rate exceeds half the inverse of the molecular relaxation time ($\dot{\varepsilon} \sim 1/2\lambda$). Under these conditions, the polymer molecules experience a transition from the coiled (equilibrium) configuration, to almost full extension.

The onset of elastic instabilities at high Wi is a hallmark of viscoelastic fluids, even under creeping flow conditions. Such purely elastic instabilities have been observed experimentally in a number of flow geometries, such as Taylor–Couette, cone-and-plate, contraction, and lid-driven cavity flows, among others [86, 121, 122]. For a thorough overview of purely elastic instabilities in (shear-dominated) viscometric flows, see the review paper by Shaqfeh [123].

Currently, it is widely accepted that the underlying mechanism for the onset of purely elastic instabilities in shear flows is related to streamline curvature, and the development of large hoop stresses, which generates tension along fluid streamlines leading to flow destabilization [86, 121, 122]. Pakdel and McKinley [86, 124] showed that the critical conditions for the onset of elastic instabilities can be described for a wide range of flows by a single dimensionless parameter, M, which accounts for elastic normal stresses and streamline curvature in the form

$$\sqrt{\frac{\lambda U \tau_{11}}{\mathfrak{R} \tau_{12}}} \equiv M \geq M_{\text{crit}} \qquad (6.36)$$

where λ is the relaxation time of the fluid, U is the local streamwise fluid velocity, τ_{11} is the local tensile stress in the flow direction, τ_{12} is the shear stress ($\tau_{12} = \eta \dot{\gamma}$), and $\mathfrak{R}$ is the streamline local radius of curvature. When the flow conditions are such that M locally exceeds a critical value, M_{crit}, elastic instabilities develop, as discussed by Pakdel and McKinley [86, 124] for several flow configurations. The value of M_{crit} is slightly dependent on the flow, and for simple flows, where the radius of curvature is known, M_{crit} can be estimated. As discussed by McKinley et al. [122], for Taylor–Couette flow $M_{\text{crit}} \approx 5.9$ and for torsional flow in a cone-and-plate arrangement, $M_{\text{crit}} \approx 4.6$. For more complex flows, the spatial variation of M needs to be taken into account to identify the critical regions where the largest value of M occurs. This mechanism for the onset of purely elastic instabilities and the applicability of the M parameter to identify the critical conditions for the onset of elastic instabilities was confirmed numerically by Alves and Poole [125] for creeping flow of UCM fluids in smooth contractions, for a wide range of contraction ratios.

6.5.3
Viscoelastic Instabilities in Some Canonical Flows

Purely elastic flow instabilities at the microscale have been observed experimentally and predicted numerically in several geometrical arrangements, such as those illustrated in Figure 6.10. The flows have been categorized in four main groups: (i) contraction/expansion flows; (ii) flows with interior stagnation points; (iii) wavy channels; (iv) other flows. In all cases, the onset of the instability can be linked to the ubiquitous presence of large normal stresses and streamline curvature in shear dominated flows (e.g., wavy channels), extensional dominated flows (e.g., stagnation/flow focusing devices), or mixed kinematic flows (e.g., contraction/expansions).

Perhaps the most widely studied configuration associated to viscoelastic fluid flow is the contraction geometry. In fact, viscoelastic flow in contraction geometries has been the subject of numerous investigations (e.g., [126, 127]). Despite relying on a simple geometrical arrangement, contraction flows usually lead to complex flow patterns, which are very sensitive to the rheological properties of the fluid, and in

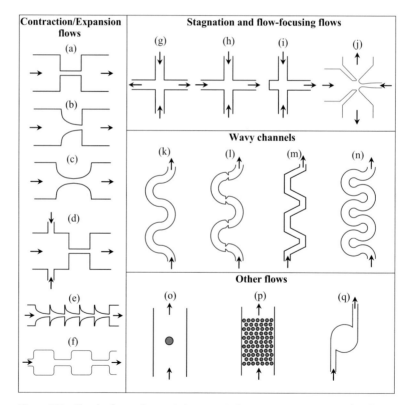

Figure 6.10 Sketch of several canonical geometrical arrangements investigated at the microscale using complex fluids that generate purely elastic flow instabilities.

particular to their extensional viscosity, geometrical details (e.g., significant differences of flow patterns are observed simply by rounding the re-entrant corner [128–130]), or the contraction ratio [131–133]. Due to their complex nature and geometrical simplicity, viscoelastic flows in abrupt contractions were established as one of the benchmark flow problems in computational rheology, during the *Vth International Workshop on Numerical Methods for Non-Newtonian Flows* [134], and since then they have been thoroughly investigated experimentally and numerically [127, 135–137]. Recent predictions of creeping flow in a 4:1 planar contraction using the Oldroyd-B model were able to reproduce the main flow features and instabilities observed experimentally in contraction flows, up to the quasi-chaotic flow observed at high Weissenberg numbers [138].

Viscoelastic flows in microscale contractions/expansions have emerged in the past decade, after the pioneering work by McKinley and co-authors [51, 139]. Microscale contraction–expansion geometries (cf. Figure 6.10a–c) enable the exploration of previously unattained regions in the Wi–Re parameter space [139], and highly elastic flow conditions can be achieved even for dilute polymer solutions as illustrated in Figure 6.3. This opens the possibility of investigating the rheology of dilute polymer solutions, particularly if hyperbolic contractions are used, as illustrated in Figure 6.10b and c, which generate a nearly constant strain rate along the centerline [61, 66]. Interestingly, the instabilities promoted at highly elastic flow conditions provide a means to enhance mixing at the microscale, as demonstrated in the experiments of Rodd *et al.* [51] for simple abrupt contraction/expansions and extensively investigated by Gan *et al.* [53, 140] and Lam *et al.* [141] who patented a modified contraction/expansion microgeometry, shown in Figure 6.10d, with additional transverse streams to trigger the instability. This microfluidic contraction/expansion device was demonstrated to work efficiently for mixing purposes at low Re and can be used with biocompatible (viscoelastic) fluids, such as polyethylene oxide (PEO) polymer solutions [142].

Viscoelastic fluid flow in contractions is usually associated with enhanced pressure drop at large Wi, when the extensional viscosity of the working fluid has a strain-hardening behavior. The different strain histories experienced in smooth contraction/abrupt expansions and abrupt contraction/smooth expansions lead to anisotropic flow resistance and can be used to develop diode-like fluidic elements, as done by Groisman and Quake [49], who used a microfluidic device consisting of a series of connected triangular elements. For the same pressure gradient applied in each direction, they achieved flow rate ratios of about 2. More recently, Sousa *et al.* [52] proposed a modified design of the microfluidic device, consisting of a series of hyperbolic elements, as in Figure 6.10e. The pressure drop in the flow direction shown in Figure 6.10e was found to be more than four times higher than the pressure drop in the opposite direction, at the same flow rate, making such a microfluidic device suitable as a fluidic equivalent of an electronic diode. The enhanced flow resistance observed in such device was found to be linked with the onset of purely elastic flow instabilities, since the corresponding purely viscous Newtonian fluid flow showed no rectification effect at these low Re. The unsteady flow of viscoelastic fluids generated at high Wi can also be used to promote efficient mixing at low Re flow conditions.

The strong extensional flow generated in microcontraction/expansions and the large strain rates that can be achieved (of about $10^5 \, s^{-1}$, or higher) make this geometrical configuration particularly interesting to study the stretching of long molecules, and in particular of DNA, under strong extensional fields. Following this idea, Gulati et al. [38] investigated the flow of semidilute solutions of λ-DNA in a 2 : 1 abrupt planar microcontraction at small Re (below 0.1) and high Wi (up to 629), corresponding to large elasticity numbers. Significant vortex enhancement was observed, particularly at high Wi, due to the highly elastic flow conditions. More recently, Hemminger et al. [143] investigated the flow of entangled DNA solutions, at different concentrations, using a 4 : 1 abrupt planar microcontraction. An unusual time-dependent shear banding flow was observed at the contraction entrance for the highest concentrations. Besides these important studies involving dilute and entangled DNA solutions, flow visualizations of the stretching and relaxation processes of individual DNA molecules in a microfluidic cross-slot geometry (cf. Figure 6.10g) have been done by Perkins et al. [144, 145], among others. The dynamics of single DNA molecules in post-arrays, as those illustrated in Figure 6.10p, have been investigated experimentally and numerically by Teclemariam et al. [146], showing that an appropriate design of post-array distribution controls DNA conformation and guides the location where the hooking events take place. A thorough review of the dynamics of a single DNA molecule in flow was presented by Shaqfeh [36].

Viscoelastic flows at high De (or high Wi) also exhibit purely elastic flow asymmetries in perfectly symmetric geometries. This steady symmetric to steady asymmetric flow transition was observed experimentally in the flow through a microscale cross-slot geometry [147] and were qualitatively captured by the 2D numerical simulations of Poole et al. [148] using the UCM model. Figure 6.11 displays a set of flow patterns predicted in the cross-slot geometry under creeping flow conditions for the UCM model for a range of De values. The Deborah number was defined as $De = \lambda U/H$ [148], where U is the average velocity on each arm of the cross-slot, with width H, as sketched in Figure 6.11. The numerical results are in qualitative agreement with the experiments of Arratia et al. [147] and show a progressive increase in the steady asymmetry above a critical Deborah number, $De_{crit} \approx 0.31$. At higher flow rates, a second instability sets in, at $De \approx 0.5$, and the flow becomes time-dependent. At significantly higher flow rates, the amplitude of oscillations increases and the flow eventually becomes chaotic, with a good mixing performance as measured by Arratia et al. [147].

Other extension-dominated flows have shown similar flow bifurcations and instabilities, as observed in the mixing–separating geometry [149], the six-arms 3D cross-slot [150], the flow-focusing device [54] (Figure 6.10h), the microfluidic T-junction geometry [55] (Figure 6.10i), or the *flip–flop* microfluidic device [48] (Figure 6.10j). For all these cases, the steady asymmetric instability occurs when large normal stresses are generated and with its onset a progressive transition from an extensionally dominated flow to a shear flow is observed to take place. This is shown in Figure 6.12 for a flow-focusing device, where the streamline patterns are superimposed onto the contour plots of the flow-type parameter, $\xi \equiv (1-R)/(1+R)$;

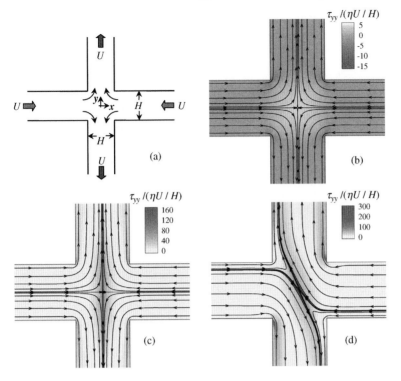

Figure 6.11 (a) Sketch of the cross-slot geometry. Streamline patterns predicted under creeping flow conditions for (b) a Newtonian fluid, and a UCM model at (c) $De = 0.3$ and (d) $De = 0.5$. The contours in (b–d) represent the normalized normal stress, $\tau_{yy}/(\eta U/H)$. Adapted from [148].

with $R = \operatorname{tr}(\mathbf{W}^2)/\operatorname{tr}(\mathbf{D}^2)$, where $\mathbf{W}$ is the relative rate of rotation tensor and $\mathbf{D}$ is the strain-rate tensor [151]. This invariant is illustrated in Figure 6.12 and varies from $\xi = -1$, corresponding to solid-like rotation flow, up to $\xi = 1$, corresponding to pure extensional flow. Shear flow corresponds to $\xi = 0$ and is easily identified near the walls, and along the channels under fully developed flow conditions.

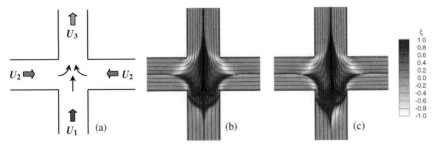

Figure 6.12 Extensional flow of a UCM fluid in a flow focusing microgeometry under creeping flow conditions. (a) Sketch of the geometry; (b) $De = 0.3$; (c) $De = 0.34$. Adapted from [54].

Despite the success in the prediction of elastic-driven steady asymmetric flow instabilities, the underlying mechanisms are yet to be fully understood, particularly the cascade of events from the first "well-behaved simple" transition to the quasi-chaotic behavior observed at very high Wi. In fact, a significantly more complex elastic instability, also not yet fully understood, is the phenomenon of elastic turbulence which also occurs for creeping flow conditions. The transition to turbulence at extremely small Re was reported for the first time by Groisman and Steinberg [88] for torsional flow of a dilute solution of a high-molecular-weight polyacrylamide between two parallel disks. In the elastic turbulence regime, despite the Reynolds number being arbitrarily small, the hallmark characteristics of classical turbulence at high Re are observed, such as enhanced flow resistance, enhanced mass and heat transfer rates, enhanced mixing, and a wide range of temporal and spatial fluctuations, as demonstrated in several subsequent experimental studies (e.g., [87, 152]), including the torsional flow between parallel plates or the flow in a wavy channel (with a square section of 3×3 mm^2, so not in the microfluidic range), as sketched in Figure 6.10k. Other investigations at the microscale involving wavy channels include the work of Groisman et al. [48] using a channel with a similar shape to that in Figure 6.10l. When using polymer solutions, such microfluidic devices work as the fluidic equivalent of a nonlinear resistor, producing a nearly constant flow rate for a wide range of pressure drops across the channel. Other studies involving zigzag channels and dilute polymer solutions showed the good mixing properties that can be achieved at high Wi, due to the onset of elastic instabilities [50] (Figure 6.10m). In contrast, for Newtonian fluids, a decrease in the mixing performance is observed when the flow rate is increased, due to the reduction in mixing time, which for low Re Newtonian flows is mainly induced by diffusion. Recently, Li et al. [153] used surfactant solutions with viscoelastic behavior (cetyltrimethyl ammonium chloride/sodium salicylate, CTAC/NaSal) and observed the onset of chaotic motion in three types of microchannels that include curved streamlines, such as wavy channels (Figure 6.10n), flow past a confined cylinder in a rectangular microchannel (Figure 6.10o), and flow in a round microfluidic cavity (Figure 6.10q). Again, the viscoelasticity of the surfactant solution together with the curved streamlines were responsible for the onset of elastic instabilities, leading to chaotic behavior and generating enhanced mixing for elastic turbulence flow conditions.

6.5.4
Elastic Turbulence

The transition to elastic turbulence depends strongly on the strain history experienced by the fluid, which is induced by the shape of the flow geometry, and on the rheological properties of the polymer solution. Nevertheless, using polymer solutions with sufficiently high elasticity, one expects that this turbulent-like motion can be excited at arbitrarily low velocities and in arbitrarily small geometries, even for very dilute polymeric solutions [154]. The elasticity of the flow increases with the inverse of the square characteristic length scale of the flow geometry (cf. Eq. (6.4)) and consequently, in microscale flows adding minute amounts of long molecules to the solution (on the order of 10 ppm, or above), is usually sufficient to induce non-Newtonian behavior at large deformation rates, which are typical of microscale flows [52, 87, 155].

So far, most of the works concerning elastic turbulence have been primarily experimental [87, 88, 152, 154, 156], and theoretical [155, 157]. The numerical simulation of elastic-driven flow instabilities has been restricted to the initial phases of flow transitions [125, 138, 148]. Only recently some preliminary attempts to simulate the elastic turbulence regime have been successful, for simplified 2D flow arrangements, such as the periodic Kolmogorov shear flow with constant forcing [158, 159]. Using direct numerical simulations (DNS) and the Oldroyd-B model to describe the fluid rheology, these authors demonstrated the occurrence of flow destabilization induced by the elastic forces due to the dynamics of polymer molecules in the solution. At large Wi, the basic phenomenology found in experimental studies of elastic turbulence was reproduced in this idealized geometrical configuration, with the appearance of coherent structures in the form of "elastic waves" [159]. Despite the use of an idealized geometrical configuration with the corresponding limitations, namely the assumption of 2D flow, Berti and Boffetta [159] demonstrated that the use of simple viscoelastic models, such as the Oldroyd-B constitutive equation, can capture the essential features of elastic turbulence, opening a window to more realistic simulations using real 3D microfluidic flow geometries and more adequate constitutive equations.

Much more needs to be investigated regarding the progressive transitions to elastic turbulence and this must be accomplished experimentally, using fluids of well-controlled rheology, and complemented with computational and theoretical studies for better insight of the complex underlying mechanisms of flow instabilities. Although there are important similarities between inertial and elastic turbulence, this does not imply that the underlying physical mechanism is the same in both cases. Indeed, elastic turbulence is accompanied by significant stretching of the polymer molecules, which is the main cause of the observed increase in the elastic normal stresses and the inherent increase in flow resistance, a ubiquitous characteristic of turbulence. The stretching of the molecular chains leads to a strong increase in flow resistance due to the increase in the extensional viscosity, a characteristic of long macromolecules in extensional flow [45], very much like the production of large Reynolds stresses in inertial turbulence of Newtonian fluids, but contrasting with the severe damping of the same Reynolds stresses that accompany polymer-induced drag reduction in high Re inertial turbulence. Understanding the nature and mechanisms that lead to elastic turbulence will have important practical applications, either for enhancing mixing and/or heat and mass transfer rates at the microscale, or for allowing the operation of extrusion processes at higher throughputs, by minimizing the driving forces that lead to the onset of flow instabilities. Additionally, understanding the driving mechanisms of elastic turbulence and comparison with classical inertial-driven turbulence of Newtonian fluids may allow us to obtain further insights into the driving mechanisms of inertial turbulence in Newtonian and in viscoelastic fluid flows.

6.6
Other Forcing Methods - Applications

The previous section discussed instabilities at high Wi flows of complex fluids driven by a pressure gradient. Here we briefly describe important works that use electrokinetic forcing to promote complex fluid flow, with emphasis on EO and electrophoresis. EOF are important in the context of viscoelastic fluids, including the development of instabilities and their possible application in micromixing enhancement. To finalize this chapter, electrophoresis is also considered, not in the framework of mixing, but rather because of its importance in the limiting case of the manipulation and separation of individual macromolecules, and of its close link with EO.

6.6.1
Electro-Osmosis

Currently, about 90% of microfluidic devices operate by either pressure-driven or EOF forcing, essentially due to their versatility and simplicity of operation [160]. PDF are still leading the number of applications of microfluidics; however as the size of the microchannels is further reduced, say to dimensions below around 10 μm, forcing by pressure becomes particularly inefficient due to the significant increase in viscous losses [160]. In contrast, for this range of dimensions EO becomes a particularly convenient and efficient way of promoting flow in microfluidic devices, as long as the fluid has ions. A major disadvantage of EOF is the strong electric gradients that typically need to be applied to promote the flow at average velocities above $1\,\mathrm{mm\,s^{-1}}$. This limitation can be circumvented by further miniaturization, thus making EOF more efficient as the size is reduced, with important applications in nanofluidics where smaller electrical potentials are sufficient to promote the flow. A thorough discussion on the advantages and disadvantages of PDF and EOF is presented in [160].

Rigorous modeling of EOF in microchannels has been the subject of several studies, particularly for Newtonian fluids. A thorough review with various applications of EO is presented in [58, 161]. Exact analytical solutions have been derived under fully developed flow conditions for Newtonian fluids, as described by Afonso et al. [162]. Newtonian fluid flow in complex geometries has been modeled in several works, and accurate results have been obtained for different applications. Of particular interest are electrokinetic instabilities (EKI) that arise under high electric fields in the presence of electrical conductivity gradients. Electrokinetic flows of Newtonian fluids become unstable when electroviscous advection of conductivity fields dominates over dissipation through viscosity and molecular diffusion [163, 164]. Likewise, EKI can be triggered using time-periodic fields, as demonstrated by Shin et al. [165] using a flow-focusing device (Figure 6.10h).

Surface patterning with different materials has been exploited to generate regions with different zeta potentials, and chaotic mixing in EOF can be driven by spatiotemporal surface charge modulation [166]. Other examples of nonlinear electrokinetic phenomena with great potential in microfluidics mixing and pumping are induced-charge electro-osmosis (ICEO) and AC electro-osmosis (ACEO), as reviewed by Bazant and Squires [167].

In contrast, EOF of complex fluids are still poorly studied, except for fully developed flows between parallel plates and in a circular tube, thus constituting a fertile ground for research. The theoretical study of EO flows of non-Newtonian fluids is recent and the preliminary works considered GNF, such as the power-law model [168, 169]). Berli and Olivares [170] considered the existence of a small wall layer depleted of additives (the skimming layer), in which the fluid behaves as a Newtonian fluid, and the non-Newtonian behavior is restricted to the electrically neutral region outside the EDL. More recently, the theoretical analysis of EOF was extended to viscoelastic fluids by Park and Lee [171], who derived the Helmholtz–Smoluchowski velocity for pure EOF of PTT fluids in rectangular channels and provided a simple numerical procedure for its calculation. Afonso et al. [162] considered the PTT and FENE-P constitutive equations, and derived analytical expressions for fully developed flow between parallel plates and in a circular pipe, under combined pressure and electrokinetic forcings. This analysis was extended by Afonso et al. [172] to consider different zeta potentials on both walls, whereas Sousa et al. [173] considered the existence of a skimming layer near the wall depleted of polymer molecules.

EOF of polymer solutions have also been studied experimentally in simple geometries. Bello et al. [174] investigated the flow of polymer solutions in capillaries, and observed a progressively suppressed EOF, suggesting a dynamic coating of the polymer molecules onto the capillary wall. Baumler et al. [175] and Chang and Tsao [176] observed drag reduction in EOF of polymer solutions, due to polymer depletion in the EDL, which leads to a reduction in shear viscosity with corresponding enhancement of the measured Helmholtz–Smoluchowski velocity.

High Weissenberg number flows are prone to purely elastic instabilities, as discussed in the previous section. EOF of polymer solutions are no exception, and Dhinakaran et al. [118] predicted a constitutive-related instability for EOF between parallel plates of PTT fluids, when the shear rate exceeds a critical value. Purely elastic instabilities can also be generated in EOF and such electroelastic instabilities were recently observed experimentally in a microfluidic channel consisting of a series of 2:1 sudden contraction/expansions (Figure 6.10f), using dilute viscoelastic PAA solutions [177, 178]. EOF are an excellent platform to generate strong extensional flows because shear effects are typically circumscribed to the EDL region. The experimental results of Bryce and Freeman [177, 178] suggest that the electroelastic instability occurs for flow conditions corresponding to the coil–stretch transition, without the observation of a dominant frequency of the flow, an indication of chaotic-like behavior. Despite this unstable behavior, the mixing rates found were smaller than those observed in polymer-free solutions, with diffusion appearing to be the dominant mixing mechanism [177]. This is a surprising result, and further investigations of electroelastic instabilities are required to enlighten the mechanism of purely elastic instabilities and mixing in microfluidic EOF of polymer solutions.

6.6.2
Electrophoresis

In many ways, DNA can be considered as an ideal model polymer. It is naturally monodisperse, large enough to be imaged in a microscope using fluorescence techniques, and its relaxation time is typically of the order of seconds [179]. DNA is a polyelectrolyte, making it easy to be manipulated through electrophoresis by applying electric fields. We note, however, that the (micro)fluidic channels used usually have charged surfaces, which also induce a global transport through EO, unless special treatments are applied to the surfaces to minimize electro-osmotic transport [180].

In a strict sense, electrophoresis is usually not included in the field of rheology. However, since it involves deformation and flow of matter, albeit in a single molecule framework, this tacit distinction is becoming obsolete, with experiments using biopolymers progressively creating an important influence on rheology [179]. Likewise, some tools developed for microrheology are expected to be increasingly used in the manipulation of single DNA molecules. A growing interaction between rheology and biophysics is leading to important insights into the flow properties of polymers and biomolecules [181]. Electrophoresis of macromolecules has several applications in molecular biology, including the transport, separation, or elongation at the molecular level, with important applications in biotechnology and medicine, with DNA sequencing being one of the most prominent. A recent review of electrophoretic microfluidic separation techniques was presented by Wu *et al.* [182], summarizing important milestones in the separation of small molecules, DNA, and proteins.

Specific microfluidic devices have been proposed and optimized in order to study individual molecules, and DNA in particular. Stretching of DNA molecules is a key technology in emerging DNA-mapping devices such as direct linear analysis [183], and single-molecule studies of DNA have expanded our knowledge on the fundamentals of polymer physics. Additionally, understanding the response of individual polymers at the molecular level can provide valuable information to develop models applicable to entangled systems. Electrophoretic stretching and relaxation of DNA molecules have been undertaken in several flow configurations that promote a strong extensional field, such as hyperbolic contractions, cross–slot, and T-junction arrangements. Juang *et al.* [184] used a cross-slot microfluidic device to induce a fairly homogeneous 2D elongational flow, allowing the determination of the amount of DNA stretching. Kim and Doyle [183] used hyperbolic contractions, which generate a nearly constant strain rate, with additional lateral streams to enhance DNA stretching. Balducci and Doyle [185] also used hyperbolic contractions, but included an obstacle array upstream of the contraction region for conformational preconditioning, leading to an increase in the average deformation of the DNA molecule in the contraction. Tang and Doyle [186] proposed a microfluidic T-shaped device which can trap and significantly stretch single DNA molecules using electrophoresis, without requiring any special end functionalization to trap the DNA molecule.

The relaxation process of DNA is also of interest as it allows for the determination of the molecular relaxation time, which is an important parameter for the characterization of DNA dynamics as shown in several experimental studies [187–189].

6.7 Conclusions and Perspectives

Flow systems built around microfluidics are becoming increasingly popular in a wide range of industrial applications dealing with both gas and liquids. They associate low production costs with low power consumption and waste reduction, allow for easy integration with electronics toward lab-on-a-chip devices, but require demanding manufacturing facilities and highly efficient signal detection systems. Many of their applications are in biotechnology and in health related areas, where the liquids are made of complex structures and macromolecules that impart nonlinear rheological behavior and in particular viscoelasticity. Since microscale flows are characterized by high surface-to-volume ratios, the flow dynamics is significantly affected by fluid rheology and other physical phenomena, such as surface tension, in comparison with macroscale flows. In particular, the time scale of the flows, $t_f \sim L/U$, decreases significantly to become much smaller than (or at least on the order of) the relaxation time of the fluid structures, leading to large Deborah number flows. As a consequence, microscale flows of complex fluids are characterized by large elastic effects in comparison to the corresponding Newtonian flows and both exhibit large ratios of viscous to inertial forces, in contrast to the corresponding macroscale flows dominated by inertia.

The large elastic effects found in microfluidic flows of complex fluids allow for enhanced mixing due to purely elastic instabilities, which are a consequence of the nonlinear nature of the corresponding terms of the rheological constitutive equation. In fact, the onset of elastic instabilities has been found to take place over a wide range of flow types (flows dominated by extension or shear, as well as flows having mixed kinematics), whenever there are large normal stresses of elastic origin coupled with streamline curvature. The dynamics of these instabilities, which exist even in the limit of creeping flow, do have some resemblance to inertial instabilities found in high Reynolds number flows of Newtonian fluids, in the sense of a progressive cascade of instabilities from simple transitions between steady flows to transitions to periodic unsteady and subsequently chaotic flows leading eventually to elastic turbulence, where the unstable flow structures exist over a continuous wide range of length and time scales. The path of these transitions is only now being discovered and remains an active topic of research. The onset of elastic turbulence in regions of parallel shear flow is currently particularly challenging.

The presence of macromolecules in microscale flow systems thus allows for enhanced mixing in low Reynolds number flows via elasticity-driven instabilities. These not only exist in pressure-gradient driven flows, but also for electrically driven flows, which are potentially very useful given their ease of implementation at these scales. The combination of EO with viscoelasticity is a new topic of research, where most things remain to be done and it suffices to think here that it is not only direct but also alternating current that has to be considered. Additionally, electrokinetic effects can be combined with surface patterning, which can also be used to enhance other surface phenomena, such as the generation of surface tension gradients. Their combination with complex fluids is in its infancy and is certainly worth exploring to find possible and unexpected flow features and applications.

Acknowledgments

The authors acknowledge funding by FEDER and Fundação para a Ciência e a Tecnologia over the last 6 years through projects REEQ/262/EME/2005, REEQ/928/EME/2005, POCI/EME/59338/2004, PTDC/EME-MFE/70186/2006, PTDC/EQU-FTT/70727/2006, PTDC/EQU-FTT/71800/2006 and PTDC/EME-MFE/099109/2008.

References

1 Squires, T.M. and Quake, S.R. (2005) Microfluidics: Fluid physics at the nanoliter scale. *Reviews of Modern Physics*, **77**, 977–1026.
2 Whitesides, G.M. (2006) The origins and the future of microfluidics. *Nature*, **442**, 368–373.
3 Saville, D.A. (1977) Electrokinetic effects with small particles. *Annual Review of Fluid Mechanics*, **9**, 321–337.
4 Probstein, R.F. (2003) *Physicochemical Hydrodynamics, an Introduction*, 2nd edn, Wiley-Interscience, Hoboken, NJ.
5 Gad-el-Hak, M.G. (2002) *The MEMS Handbook*, CRC Press, Boca Raton, FL.
6 Bruus, H. (2008) *Theoretical Microfluidics, Oxford Master Series in Condensed Matter Physics*, Oxford University Press, Oxford, UK.
7 Ajdari, A. (2000) Pumping liquids using asymmetric electrode arrays. *Physical Review E*, **61**, R45–R48.
8 Bazant, M.Z. and Squires, T.M. (2004) Induced-charge electrokinetic phenomena: Theory and microfluidic applications. *Physical Review Letters*, **92**, 066101.
9 Stone, H.A., Stroock, A.D., and Ajdari, A. (2004) Engineering flows in small devices: microfluidics toward lab-on-a-chip. *Annual Review of Fluid Mechanics*, **36**, 381–411.
10 Laser, D.J. and Santiago, J.G. (2004) A review of micropumps. *Journal of Micromechanics and Microengineering*, **14**, R35–R64.
11 Stone, H.A. (2009) Microfluidics, tuned-in flow control. *Nature Physics*, **5**, 178–179.
12 Brody, J.P., Kamholz, A.E., and Yager, P. (1997) Prominent microscopic effects in microfabricated fluidic analysis systems. Micro- and Nanofabricated Electro-Optical Mechanical Systems for Biomedical and Environmental Applications, Proceedings of, **2978** pp. 103–110.
13 Hatch, A., Kamholz, A.E., Hawkins, K.R., Munson, M.S., Schilling, E.A., Weigl, B.H., and Yager, P. (2001) A rapid diffusion immunoassay in a T-sensor. *Nature Biotechnology*, **19**, 461–465.
14 Dendukuri, D., Tsoi, K., Hatton, T.A., and Doyle, P.S. (2005) Controlled synthesis of nonspherical microparticles using microfluidics. *Langmuir*, **21**, 2113–2116.
15 Nguyen, N.-T. (2008) *Micromixers: Fundamentals, Design and Fabrication*, William Andrew, Norwich, NY.
16 Sia, S.K. and Whitesides, G.M. (2003) Microfluidic devices fabricated in poly (dimethylsiloxane) for biological studies. *Electrophoresis*, **24**, 3563–3576.

17 Pittet, P., Lu, G.N., Galvan, J.M., Ferrigno, R., Stephan, K., Blum, L.J., and Leca-Bouvier, B. (2008) A novel low-cost approach of implementing electrochemiluminescence detection for microfluidic analytical systems. *Materials Science & Engineering C-Biomimetic and Supramolecular Systems*, **28**, 891–895.

18 Kallio, P. and Kuncova, J. (2004) Microfluidics. *Technology Review*, **158**, 2004.

19 Nguyen, N.T. and Wu, Z.G. (2005) Micromixers – a review. *Journal of Micromechanics and Microengineering*, **15**, R1–R16.

20 Oh, K.W. and Ahn, C.H. (2006) A review of microvalves. *Journal of Micromechanics and Microengineering*, **16**, R13–R39.

21 Quake, S.R. and Scherer, A. (2000) From micro- to nanofabrication with soft materials. *Science*, **290**, 1536–1540.

22 Ng, J.M.K., Gitlin, I., Stroock, A.D., and Whitesides, G.M. (2002) Components for integrated poly(dimethylsiloxane) microfluidic systems. *Electrophoresis*, **23**, 3461–3473.

23 Marrian, C.R.K. and Tennant, D.M. (2003) Nanofabrication. *Journal of Vacuum Science & Technology A – Vacuum Surfaces and Films*, **21**, S207–S215.

24 Bayraktar, T. and Pidugu, S.B. (2006) Characterization of liquid flows in microfluidic systems. *International Journal of Heat and Mass Transfer*, **49**, 815–824.

25 Whitesides, G.M. and Stroock, A.D. (2001) Flexible methods for microfluidics. *Physics Today*, **54**, 42–48.

26 Thorsen, T., Roberts, R.W., Arnold, F.H., and Quake, S.R. (2001) Dynamic pattern formation in a vesicle-generating microfluidic device. *Physical Review Letters*, **86**, 4163–4166.

27 Anna, S.L., Bontoux, N., and Stone, H.A. (2003) Formation of dispersions using "flow focusing" in microchannels. *Applied Physics Letters*, **82**, 364–366.

28 Yang, T.L., Jung, S.Y., Mao, H.B., and Cremer, P.S. (2001) Fabrication of phospholipid bilayer-coated microchannels for on-chip immunoassays. *Analytical Chemistry*, **73**, 165–169.

29 Fu, A.Y., Spence, C., Scherer, A., Arnold, F.H., and Quake, S.R. (1999) A microfabricated fluorescence-activated cell sorter. *Nature Biotechnology*, **17**, 1109–1111.

30 Sohn, L.L., Saleh, O.A., Facer, G.R., Beavis, A.J., Allan, R.S., and Notterman, D.A. (2000) Capacitance cytometry: Measuring biological cells one by one. *Proceedings of the National Academy of Sciences of the United States of America*, **97**, 10687–10690.

31 Chiu, D.T., Jeon, N.L., Huang, S., Kane, R.S., Wargo, C.J., Choi, I.S., Ingber, D.E., and Whitesides, G.M. (2000) Patterned deposition of cells and proteins onto surfaces by using three-dimensional microfluidic systems. *Proceedings of the National Academy of Sciences of the United States of America*, **97**, 2408–2413.

32 Li, J.J., Kelly, J.F., Chemushevich, I., Harrison, D.J., and Thibault, P. (2000) Separation and identification of peptides from gel-isolated membrane proteins using a microfabricated device for combined capillary electrophoresis/nanoelectrospray mass spectrometry. *Analytical Chemistry*, **72**, 599–609.

33 Beebe, D.J., Mensing, G.A., and Walker, G.M. (2002) Physics and applications of microfluidics in biology. *Annual Review of Biomedical Engineering*, **4**, 261–286.

34 Zare, R.N. and Kim, S. (2010) Microfluidic platforms for single-cell analysis. *Annual Review of Biomedical Engineering*, **12**, 187–201.

35 Tegenfeldt, J.O., Prinz, C., Cao, H., Huang, R.L., Austin, R.H., Chou, S.Y., Cox, E.C., and Sturm, J.C. (2004) Micro- and nanofluidics for DNA analysis. *Analytical and Bioanalytical Chemistry*, **378**, 1678–1692.

36 Shaqfeh, E.S.G. (2005) The dynamics of single-molecule DNA in flow. *Journal of Non-Newtonian Fluid Mechanics*, **130**, 1–28.

37 Randall, G.C., Schultz, K.M., and Doyle, P.S. (2006) Methods to electrophoretically stretch DNA: Microcontractions, gels, and hybrid gel-microcontraction devices. *Lab on a Chip*, **6**, 516–525.

38 Gulati, S., Muller, S.J., and Liepmann, D. (2008) Direct measurements of viscoelastic flows of DNA in a 2:1 abrupt planar micro-contraction. *Journal of Non-Newtonian Fluid Mechanics*, **155**, 51–66.

39 Choban, E.R., Markoski, L.J., Wieckowski, A., and Kenis, P.J.A. (2004) Microfluidic fuel cell based on laminar flow. *Journal of Power Sources*, **128**, 54–60.

40 Obot, N.T. (2002) Toward a better understanding of friction and heat/mass transfer in microchannels – A literature review. *Microscale Thermophysical Engineering*, **6**, 155–173.

41 Hansen, C. and Quake, S.R. (2003) Microfluidics in structural biology: smaller, faster... better. *Current Opinion in Structural Biology*, **13**, 538–544.

42 Gad-el-Hak, M. (2005) Liquids: The holy grail of microfluidic modeling. *Physics of Fluids*, **17**, 100612.

43 Breussin, F. (2009) Emerging markets for microfluidic applications in life sciences and in-vitro diagnostics, EMMA Report, Yole Développment SA.

44 Tabeling, P. (2005) *Introduction to Microfluidics*, Oxford University Press, Oxford.

45 Bird, R.B., Armstrong, R.C., and Hassager, O. (1987) *Dynamics of Polymeric Liquids. Volume 1: Fluid Mechanics*, Wiley, New York.

46 Larson, R.G. (1999) *The Structure and Rheology of Complex Fluids*, Oxford University Press, Oxford.

47 Dealy, J.M. (2010) Weissenberg and Deborah numbers – Their definition and use. *Rheology Bulletin*, **79**, 14–18.

48 Groisman, A., Enzelberger, M., and Quake, S.R. (2003) Microfluidic memory and control devices. *Science*, **300**, 955–958.

49 Groisman, A. and Quake, S.R. (2004) A microfluidic rectifier: anisotropic flow resistance at low Reynolds numbers. *Physical Review Letters*, **92**, 094501-1–094501-4.

50 Pathak, J.A., Ross, D., and Migler, K.B. (2004) Elastic flow instability, curved streamlines, and mixing in microfluidic flows. *Physics of Fluids*, **16**, 4028-1–4028-7.

51 Rodd, L.E., Scott, T.P., Boger, D.V., Cooper-White, J.J., and McKinley, G.H. (2005) The inertio-elastic planar entry flow of low-viscosity elastic fluids in micro-fabricated geometries. *Journal of Non-Newtonian Fluid Mechanics*, **129**, 1–22.

52 Sousa, P.C., Pinho, F.T., Oliveira, M.S.N., and Alves, M.A. (2010) Efficient microfluidic rectifiers for viscoelastic fluid flow. *Journal of Non-Newtonian Fluid Mechanics*, **165**, 652–671.

53 Gan, H.Y., Lam, Y.C., and Nguyen, N.T. (2006) Polymer-based device for efficient mixing of viscoelastic fluids. *Applied Physics Letters*, **88**, 224103.

54 Oliveira, M.S.N., Pinho, F.T., Poole, R.J., Oliveira, P.J., and Alves, M.A. (2009) Purely-elastic flow asymmetries in flow-focusing devices. *Journal of Non-Newtonian Fluid Mechanics*, **160**, 31–39.

55 Soulages, J., Oliveira, M.S.N., Sousa, P.C., Alves, M.A., and McKinley, G.H. (2009) Investigating the stability of viscoelastic stagnation flows in T-shaped microchannels. *Journal of Non-Newtonian Fluid Mechanics*, **163**, 9–24.

56 Batchelor, G.K. (2000) *An Introduction to Fluid Dynamics*, Cambridge University Press, Cambridge, UK; New York, NY.

57 Nguyen, N.-T. and Wereley, S.T. (2006) *Fundamentals and Applications of Microfluidics*, 2nd edn, Artech House, Norwood.

58 Karniadakis, G., Beskok, A., and Aluru, N. (2005) *Microflows and nanoflows. Fundamentals and Simulation*, Interdisciplinary Applied Mathematics Series, vol. 29, Springer, Berlin.

59 Koo, J.M. and Kleinstreuer, C. (2003) Liquid flow in microchannels: experimental observations and computational analyses of microfluidics effects. *Journal of Micromechanics and Microengineering*, **13**, 568–579.

60 Sharp, K.V. and Adrian, R.J. (2004) Transition from laminar to turbulent flow in liquid filled microtubes. *Exp Fluids*, **36**, 741–747.

61 Oliveira, M.S.N., Alves, M.A., Pinho, F.T., and McKinley, G.H. (2007) Viscous flow through microfabricated hyperbolic

contractions. *Experiments in Fluids*, **43**, 437–451.

62 Oliveira, M.S.N., Rodd, L.E., McKinley, G.H., and Alves, M.A. (2008) Simulations of extensional flow in microrheometric devices. *Microfluid Nanofluid*, **5**, 809–826.

63 Pit, R., Hervet, H., and Leger, L. (2000) Direct experimental evidence of slip in hexadecane: Solid interfaces. *Physical Review Letters*, **85**, 980–983.

64 Barrat, J.L. and Bocquet, L. (1999) Large slip effect at a nonwetting fluid-solid interface. *Physical Review Letters*, **82**, 4671–4674.

65 Colin, S. (2005) Rarefaction and compressibility effects on steady and transient gas flows in microchannels. *Microfluid Nanofluid*, **1**, 268–279.

66 Pipe, C.J. and McKinley, G.H. (2009) Microfluidic rheometry. *Mechanics Research Communications*, **36**, 110–120.

67 Ottino, J.M. and Wiggins, S. (2004) Applied physics – Designing optimal micromixers. *Science*, **305**, 485–486.

68 Wang, H.Z., Iovenitti, P., Harvey, E., and Masood, S. (2002) Optimizing layout of obstacles for enhanced mixing m microchannels. *Smart Materials & Structures*, **11**, 662–667.

69 Brody, J.P., Yager, P., Goldstein, R.E., and Austin, RH. (1996) Biotechnology at low Reynolds numbers. *Biophysical Journal*, **71**, 3430–3441.

70 Ottino, J.M. (2004) *The Kinematics of Mixing: Stretching, Chaos, and Transport*, Cambridge University Press, Cambridge, UK.

71 Coleman, J.T. and Sinton, D. (2005) A sequential-injection microfluidic mixing strategy. *Microfluid Nanofluid*, **1**, 319–327.

72 Glasgow, I. and Aubry, N. (2003) Enhancement of microfluidic mixing using time pulsing. *Lab on a Chip*, **3**, 114–120.

73 Deshmukh, A.A., Liepmann, D., and Pisano, A.P. (2000) Continuous micromixer with pulsatile micropumps. Technical Digest of the IEEE Solid State Sensor and Actuator Workshop (Hilton Nead Island, SC) 73–76.

74 El Moctar, A.O., Aubry, N., and Batton, J. (2003) Electro-hydrodynamic microfluidic mixer. *Lab on a Chip*, **3**, 273–280.

75 Bau, H.H., Zhong, J., and Yi, M. (2001) A minute magneto hydro dynamic (MHD) mixer. *Sensors and Actuators B: Chemical*, **79**, 207–215.

76 Qian, S. and Bau, H.H. (2009) Magneto-hydrodynamics based microfluidics. *Mechanics Research Communications*, **36**, 10–21.

77 Yang, Z., Goto, H., Matsumoto, M., and Maeda, R. (2000) Active micromixer for microfluidic systems using lead-zirconate-titanate (PZT)-generated ultrasonic vibration. *Electrophoresis*, **21**, 116–119.

78 Yang, Z., Matsumoto, S., Goto, H., Matsumoto, M., and Maeda, R. (2001) Ultrasonic micromixer for microfluidic systems. *Sensor Actuators A – Phys*, **93**, 266–272.

79 Hessel, V., Lowe, H., and Schonfeld, F. (2005) Micromixers – a review on passive and active mixing principles. *Chemical Engineering Science*, **60**, 2479–2501.

80 Aref, H. (1984) Stirring by chaotic advection. *Journal of Fluid Mechanics*, **143**, 1–21.

81 Stroock, A.D., Dertinger, S.K.W., Ajdari, A., Mezic, I., Stone, H.A., and Whitesides, G.M. (2002) Chaotic mixer for microchannels. *Science*, **295**, 647–651.

82 Gleeson, J.P. (2005) Transient micromixing: Examples of laminar and chaotic stirring. *Physics of Fluids*, **17**, 100614.

83 Tadmor, Z. and Gogos, C.G. (2006) *Principles of Polymer Processing*, 2nd edn, Wiley-Interscience, Hoboken, NJ.

84 Oliveira, M.S.N., Pinho, F.T., and Alves, M.A. (2008) Extensional effects in viscoelastic fluid flow through a microscale double cross-slot. Xvth International Congress on Rheology – the Society of Rheology 80th Annual Meeting, Pts 1 and 2, 1027, pp. 982–984.

85 Townsend, P., Walters, K., and Waterhouse, W.M. (1976) Secondary flows in pipes of square cross-section and measurement of 2nd normal stress difference. *Journal of Non-Newtonian Fluid Mechanics*, **1**, 107–123.

86 Pakdel, P. and McKinley, G.H. (1996) Elastic instability and curved streamlines. *Physical Review Letters*, **77**, 2459–2462.

87 Groisman, A. and Steinberg, V. (2004) Elastic turbulence in curvilinear flows of polymer solutions. *New Journal of Physics*, **6-29**, 1–48.

88 Groisman, A. and Steinberg, V. (2000) Elastic turbulence in a polymer solution flow. *Nature*, **405**, 53–55.

89 Whorlow, R.W. (1992) *Rheological Techniques*, 2nd edn, Ellis Horwood Series in Physics and Its Applications, London.

90 Walters, K. (1975) *Rheometry*, Chapman and Hall, London.

91 Chen, D.C.-H. (1986) Yield stress: A time-dependent property and how to measure it. *Rheologica Acta*, **25**, 542–554.

92 Nguyen, Q.D. and Boger, D.V. (1992) Measuring the flow properties of yield stress fluids. *Annual Review of Fluid Mechanics*, **24**, 47–88.

93 Barnes, H.A. and Walters, K. (1985) The yield stress myth? *Rheologica Acta*, **24**, 323–326.

94 Barnes, H.A. (1995) A review of the slip (wall depletion) of polymer solutions, emulsions and particle suspensions in viscometers: its cause, character and cure. *Journal of Non-Newtonian Fluid Mechanics*, **56**, 221–251.

95 Barnes, H.A. (1999) The yield stress – a review or 'παντα ρει'– everything flows? *Journal of Non-Newtonian Fluid Mechanics*, **81**, 133–178.

96 Barnes, H.A., Hutton, J.F., and Walters, K. (1989) *An Introduction to Rheology*, Elsevier, Amsterdam.

97 Alves, M.A., Pinho, F.T., and Oliveira, P.J. (2005) Visualizations of Boger fluid flows in a 4:1 square–square contraction. *AICHE Journal*, **51**, 2908–2922.

98 Rodd, L.E., Scott, T.P., Cooper-White, J.J., and McKinley, G.H. (2005) Capillary breakup rheometry for low viscosity elastic fluids. *Applied Rheology*, **15**, 12–27.

99 Tirtaatmadja, V. and Sridhar, T. (1993) A filament stretching device for measurement of extensional viscosity. *Journal of Rheology*, **37**, 1081–1102.

100 McKinley, G.H. and Sridhar, T. (2002) Filament-stretching rheometry of complex fluids. *Annual Review of Fluid Mechanics*, **34**, 375–415.

101 Sentmanat, M.L. (2003) A novel device for characterizing polymer flows in uniaxial extension. SPE, ANTEC Proceedings, pp. 992–996.

102 Lielens, G., Keunings, R., and Legat, V. (1999) The FENE-L and FENE-LS closure approximations to the kinetic theory of finitely extensible dumbbells. *Journal of Non-Newtonian Fluid Mechanics*, **87**, 179–196.

103 Tanner, R.I. (2000) *Engineeering Rheology*, 2nd edn, Oxford Eng. Science Series 52, Oxford University Press, New York.

104 Ewoldt, R.H., Hosoi, A.E., and McKinley, G.H. (2008) New measures for characterizing nonlinear viscoelasticity in large amplitude oscillatory shear. *Journal of Rheology*, **35** (4), 647–685.

105 Beris, A.N. and Edwards, B.J. (1994) *Thermodynamics of Flowing Systems*, Oxford Engineering Science Series 36, Oxford Science Publications, New York.

106 Wapperom, P. (1995) Nonisothermal flows of viscoelastic fluids. Thermodynamics, analysis and numerics. PhD thesis. Technical University of Delft, Holland.

107 Wapperom, P. and Hulsen, M.A. (1998) Thermodynamics of viscoelastic fluids: the temperature equation. *Journal of Rheology*, **42**, 999–1019.

108 Nóbrega, J.M., Pinho, F.T., Oliveira, P.J., and Carneiro, O.S. (2004) Accounting for temperature-dependent properties in viscoelastic duct flows. *International Journal of Heat and Mass Transfer*, **47**, 1141–1158.

109 Bird, R.B., Armstrong, R.C., and Hassager, O. (1987) *Dynamics of Polymeric Liquids. Volume 2: Kinetic Theory*, Wiley, New York.

110 Bird, R.B., Stewart, W.E., and Lightfoot, E.N. (2002) *Transport Phenomena*, 2nd edn, John Wiley & Sons Inc., New York.

111 Astarita, G. and Marrucci, G. (1974) *Principles of Non-Newtonian Fluid Mechanics*, McGraw-Hill, London, New York.

112 Ferry, J.D. (1981) *Viscoelastic Properties of Polymers*, 2nd edn, John Wiley, New York.

113 Wapperom, P., Hulsen, M.A., and van der Zanden, J.P.P.M. (1998) A numerical method for steady and nonisothermal viscoelastic fluid flow for high Deborah and Péclet numbers. *Rheologica Acta*, **37**, 73–88.

114 Chilcott, M.D. and Rallison, J.M. (1988) Creeping flow of dilute polymer solutions past cylinders and spheres. *Journal of Non-Newtonian Fluid Mechanics*, **29**, 381–432.

115 Stokes, J.R., Graham, L.J.W., Lawson, N.J., and Boger, D.V. (2001) Swirling flow of viscoelastic fluids. Part 2: Elastic effects. *Journal of Fluid Mechanics*, **429**, 117–153.

116 Larson, R.G. (1988) *Constitutive Equations for Polymer Melts and Solutions*, Butterworths, Boston.

117 Huilgol, R.R. and Phan-Thien, N. (1997) *Fluid Mechanics of Viscoelasticity*, Elsevier, Amsterdam.

118 Dhinakaran, S., Afonso, A.M., Alves, M.A., and Pinho, F.T. (2010) Steady viscoelastic fluid flow in microchannels under electrokinetic forces: PTT model. *Journal of Colloid and Interface Science*, **344**, 513–520.

119 Bird, R.B. and Curtiss, C.F. (1998) Thermoviscoelasticity: continuum-molecular connections. *Journal of Non-Newtonian Fluid Mechanics*, **79**, 255–259.

120 Peters, G.W.M. and Baaijens, F.P.T. (1997) Modelling of non-isothermal viscoelastic flows. *Journal of Non-Newtonian Fluid Mechanics*, **68**, 205–224.

121 Larson, R.G., Shaqfeh, E.S.G., and Muller, S.J. (1990) A purely elastic instability in Taylor–Couette flow. *Journal of Fluid Mechanics*, **218**, 573–600.

122 McKinley, G.H., Pakdel, P., and Öztekin, A. (1996) Rheological and geometric scaling of purely elastic flow instabilities. *Journal of Non-Newtonian Fluid Mechanics*, **67**, 19–47.

123 Shaqfeh, E.S.G. (1996) Purely elastic instabilities in viscometric flows. *Annual Review of Fluid Mechanics*, **28**, 129–185.

124 Pakdel, P. and McKinley, G.H. (1998) Cavity flows of elastic liquids: Purely elastic instabilities. *Physics of Fluids*, **10**, 1058–1070.

125 Alves, M.A. and Poole, R.J. (2007) Divergent flow in contractions. *Journal of Non-Newtonian Fluid Mechanics*, **144**, 140–148.

126 Boger, D.V. (1987) Viscoelastic flows through contractions. *Annual Review of Fluid Mechanics*, **19**, 157–182.

127 Owens, R.G. and Phillips, T.N. (2002) *Computational Rheology*, Imperial College Press, London.

128 Evans, R.E. and Walters, K. (1986) Flow characteristics associated with abrupt changes in geometry in the case of highly elastic liquids. *Journal of Non-Newtonian Fluid Mechanics*, **20**, 11–29.

129 Evans, R.E. and Walters, K. (1989) Further remarks on the lip-vortex mechanism of vortex enhancement in planar-contraction flows. *Journal of Non-Newtonian Fluid Mechanics*, **32**, 95–105.

130 Rothstein, J.P. and McKinley, G.H. (1999) Extensional flow of a polystyrene Boger fluid through a 4:1:4 axisymmetric contraction-expansion. *Journal of Non-Newtonian Fluid Mechanics*, **86**, 61–88.

131 Boger, D.V., Hur, D.U., and Binnington, R.J. (1986) Further observations of elastic effects in tubular entry flows. *Journal of Non-Newtonian Fluid Mechanics*, **20**, 31–49.

132 Alves, M.A., Oliveira, P.J., and Pinho, F.T. (2004) On the effect of contraction ratio in viscoelastic flow through abrupt contractions. *Journal of Non-Newtonian Fluid Mechanics*, **122**, 117–130.

133 Oliveira, M.S.N., Pinho, F.T., Oliveira, P.J., and Alves, M.A. (2007) Effect of contraction ratio upon viscoelastic flow in contractions: The axisymmetric case. *Journal of Non-Newtonian Fluid Mechanics*, **47**, 92–108.

134 Hassager, O. (1988) Working group on numerical techniques. in: Proceedings of the Vth Workshop on numerical methods in non-Newtonian flow. *Journal of Non-Newtonian Fluid Mechanics*, **29**, 2–5.

135 Alves, M.A., Oliveira, P.J., and Pinho, F.T. (2003) Benchmark solutions for the flow of Oldroyd-B and PTT fluids in planar contractions. *Journal of Non-Newtonian Fluid Mechanics*, **110**, 45–75.

136 Walters, K. and Webster, M.F. (2003) The distinctive CFD challenges of

computational rheology. *International Journal for Numerical Methods in Fluids*, **43**, 577–596.

137 Kim, J.M., Kim, C., Kim, J.H., Chung, C., Ahn, K.H., and Lee, S.J. (2005) High-resolution finite element simulation of 4:1 planar contraction flow of viscoelastic fluid. *Journal of Non-Newtonian Fluid Mechanics*, **129**, 23–37.

138 Afonso, A., Oliveira, P.J., Pinho, F.T., and Alves, M.A. (2011) Dynamics of high Deborah number entry flows: A numerical study. *Journal of Fluid Mechanics*, **677**, 272–304.

139 Rodd, L.E., Cooper-White, J.J., McKinley, G.H., and Boger, D.V. (2007) Role of the elasticity number in entry flow of dilute polymer solutions in microfabricated contraction geometries. *Journal of Non-Newtonian Fluid Mechanics*, **143**, 170–191.

140 Gan, H.Y., Lam, Y.C., Nguyen, N.T., Tam, K.C., and Yang, C. (2007) Efficient mixing of viscoelastic fluids in a microchannel at low Reynolds number. *Microfluid Nanofluid*, **3**, 101–108.

141 Lam, Y.C., Gan, H.Y., Nguyen, N.T., and Lie, H. (2009) Micromixer based on viscoelastic flow instability at low Reynolds number. *Biomicrofluidics*, **3**, 014106.

142 Lam, Y.C., Gan, H.Y., Nguyen, N.T., Yang, C., and Tam, K.C. (2008) Methods and apparatus for microfluidic mixing, US Patent, 2008/0259720.

143 Hemminger, O.L., Boukany, P.E., Wang, S.-Q.-., and Lee, L.J. (2010) Flow pattern and molecular visualization of DNA solutions through a 4:1 planar micro-contraction. *Journal of Non-Newtonian Fluid Mechanics*, **165**, 1613–1624.

144 Perkins, T.T., Smith, D.E., Larson, R.G., and Chu, S. (1995) Stretching of a single tethered polymer in a uniform flow. *Science*, **268**, 83–87.

145 Perkins, T.T., Smith, D.E., and Chu, S. (1997) Single polymer dynamics in an elongational flow. *Science*, **276**, 2016–2021.

146 Teclemariam, N.P., Beck, V.A., Shaqfeh, E.S.G., and Muller, S.J. (2007) Dynamics of DNA polymers in post arrays: Comparison of single molecule experiments and simulations. *Macromolecules*, **40**, 3848–3859.

147 Arratia, P.E., Thomas, C.C., Diorio, J., and Gollub, J.P. (2006) Elastic instabilities of polymer solutions in cross-channel flow. *Physical Review Letters*, **96**, 144502-1–144502-4.

148 Poole, R.J., Alves, M.A., and Oliveira, P.J. (2007) Purely-elastic flow asymmetries. *Physical Review Letters*, **99**, 164503.

149 Afonso AM, Alves MA, Poole RJ, Oliveira PJ and Pinho FT 2011. Viscoelastic flows in mixing-separating cells. *Journal of Engineering Mathematics*, Special Issue on Complex Flows, **71**, 3–13.

150 Afonso, A.M., Alves, M.A., and Pinho, F.T. (2010) Purely-elastic flow instabilities in a 3D six arms cross slot geometry. *Journal of Non-Newtonian Fluid Mechanics*, **165**, 743–751.

151 Mompean, G., Thompson, R.L., and Mendes, P.R.S. (2003) A general transformation procedure for differential viscoelastic models. *Journal of Non-Newtonian Fluid Mechanics*, **111**, 151–174.

152 Schiamberg, B.A., Shereda, L.T., Hu, H., and Larson, R.G. (2006) Transitional pathway to elastic turbulence in torsional, parallel-plate flow of a polymer solution. *Journal of Fluid Mechanics*, **554**, 191–216.

153 Li, F.-C., Kinoshita, H., Li, X.-B., Oishi, M., Fujii, T., and Oshima, M. (2010) Creation of very-low-Reynolds-number chaotic fluid motions in microchannels using viscoelastic surfactant solutions. *Experimental Thermal and Fluid Science*, **34**, 20–27.

154 Burghelea, T., Segre, E., and Steinberg, V. (2007) Elastic turbulence in von Kármán swirling flow between two disks. *Physics of Fluids*, **19**, 053104.

155 Fouxon, A. and Lebedev, V. (2007) Spectra of turbulence in dilute polymer solutions. *Physics of Fluids*, **15**, 2060–2072.

156 Fardin, M.A., Lopez, D., Croso, J., Grégoire, G., Cardoso, O., McKinley, G.H., and Lerouge, S. (2010) Elastic turbulence in shear banding wormlike micelles. *Physical Review Letters*, **104**, 178303.

157 Morozov, A.N. and van Saarloos, W. (2007) An introductory essay on subcritical instabilities and the transition to turbulence in viscoelastic parallel shear flows. *Psychological Reports*, **447**, 112–143.

158 Bistagnino, A., Boffetta, G., Celani, A., Mazzino, A., Puliafito, A., and Vergassola, M. (2007) Nonlinear dynamics of the viscoelastic Kolmogorov flow. *Journal of Fluid Mechanics*, **590**, 61–80.

159 Berti, S. and Boffetta, G. (2010) Elastic waves and transition to elastic turbulence in a two-dimensional viscoelastic Kolmogorov flow. *Physical Review E*, **82**, 036314.

160 Pennathur, S. (2008) Flow control in microfluidics: Are the workhorse flows adequate? *Lab on a Chip*, **8**, 383–387.

161 Li, D. (ed.) (2008) *Encyclopedia of Microfluidic and Nanofluidic*, Springer, Berlin.

162 Afonso, A.M., Alves, M.A., and Pinho, F.T. (2009) Analytical solution of mixed electro-osmotic/pressure driven flows of viscoelastic fluids in microchannels. *Journal of Non-Newtonian Fluid Mechanics*, **159**, 50–63.

163 Lin, H., Storey, B.D., Oddy, M.H., Chen, C.H., and Santiago, J.G. (2004) Instability of electrokinetic microchannel flows with conductivity gradients. *Physics of Fluids*, **16**, 1922–1935.

164 Chen, C.-H., Lin, H., Lele, S.K., and Santiago, J.G. (2005) Convective and absolute electrokinetic instability with conductivity gradients. *Journal of Fluid Mechanics*, **524**, 263–303.

165 Shin, S.M., Kang, I.S., and Cho, Y.-K. (2005) Mixing enhancement by using electrokinetic instability under time-periodic electric field. *Journal of Micromechanics and Microengineering*, **15**, 455–462.

166 Chang, C.-C. and Yang, R.-J. (2009) Chaotic mixing in electro-osmotic flows driven by spatiotemporal surface charge modulation. *Physics of Fluids*, **21**, 052004.

167 Bazant, M.Z. and Squires, T.M. (2010) Induced-charge electrokinetic phenomena. *Current Opinion in Colloid and Interface Science*, **15**, 203–213.

168 Das, S. and Chakraborty, S. (2006) Analytical solutions for velocity, temperature and concentration distribution in electroosmotic microchannel flows of a non-Newtonian bio-fluid. *Analytica Chimica Acta*, **559**, 15–24.

169 Chakraborty, S. (2007) Electroosmotically driven capillary transport of typical non-Newtonian biofluids in rectangular microchannels. *Analytica Chimica Acta*, **605**, 175–184.

170 Berli, C.L.A. and Olivares, M.L. (2008) Electrokinetic flow of non-Newtonian fluids in microchannels. *Journal of Colloid and Interface Science*, **320**, 582–589.

171 Park, H.M. and Lee, W.M. (2008) Helmholtz-Smoluchowski velocity for viscoelastic electroosmotic flows. *Journal of Colloid and Interface Science*, **317**, 631–636.

172 Afonso AM, Alves MA and Pinho FT 2011. Electro-osmotic flow of viscoelastic fluids in microchannels under asymmetric zeta potentials. *Journal of Engineering Mathematics*, **71**, 15–30.

173 Sousa, J.J., Afonso, A.M., Pinho, F.T., and Alves, M.A. (2011) Effect of the skimming layer on electro-osmotic-Poiseuille flows of viscoelastic fluids. *Microfluid Nanofluid*, **10**, 107–122.

174 Bello, M.S., De Besi, P., Rezzonico, R., Righetti, P.G., and Casiraghi, E. (1994) Electroosmosis of polymer solutions in fused silica capillaries. *Electrophoresis*, **15**, 623–626.

175 Baumler, H., Neu, B., Iovtchev, S., Budde, A., Kiesewetter, H., Latza, R., and Donath, E. (1999) Electroosmosis and polymer depletion layers near surface conducting particles are detectable by low frequency electrorotation. *Colloids and Surfaces A*, **149**, 389–396.

176 Chang, F.-M. and Tsao, H.-K. (2007) Drag reduction in electro-osmosis of polymer solutions. *Applied Physics Letters*, **90**, 194105.

177 Bryce, R.M. and Freeman, M.R. (2010) Abatement of mixing in shear-free elongationally unstable viscoelastic microflows. *Lab on a Chip*, **10**, 1436–1441.

178 Bryce, R.M. and Freeman, M.R. (2010) Extensional instability in electro-osmotic microflows of polymer solutions. *Physical Review E*, **81**, 036328.

179 Graham, R.S. and McLeish, T.C.B. (2008) Emerging applications for models of molecular rheology. *Journal of Non-Newtonian Fluid Mechanics*, **150**, 11–18.

180 Hsieh, C.-C. and Doyle, P.S. (2008) Studying confined polymers using single-molecule DNA experiments. *Korea-Australia Rheology Journal*, **20**, 127–142.

181 Larson, R.G. (2007) Going with the flow. *Science*, **318**, 57–58.

182 Wu, D., Qin, J., and Lin, B. (2008) Electrophoretic separations on microfluidic chips. *Journal of Chromatography. A*, **1184**, 542–559.

183 Kim, J.M. and Doyle, P.S. (2007) Design and numerical simulation of a DNA electrophoretic stretching device. *Lab on a Chip*, **7**, 213–225.

184 Juang, Y.-J., Wang, S., Hu, X., and Lee, L.J. (2004) Dynamics of single polymers in a stagnation flow induced by electrokinetics. *Physical Review Letters*, **93**, 268105.

185 Balducci, A. and Doyle, P.S. (2008) Conformational preconditioning by electrophoresis of DNA through a finite obstacle array. *Macromolecules*, **41**, 5485–5492.

186 Tang, J. and Doyle, P.S. (2007) Electrophoretic stretching of DNA molecules using microscale T junctions. *Applied Physics Letters*, **90**, 224103.

187 Perkins, T.T., Quake, S.R., Smith, D.E., and Chu, S. (1994) Relaxation of a single DNA molecule observed by optical microscopy. *Science*, **264**, 822–826.

188 Ferree, S. and Blanch, H.W. (2004) The hydrodynamics of DNA electrophoretic stretch and relaxation in a polymer solution. *Biophysical Journal*, **87**, 468–475.

189 Liu, Y., Jun, Y., and Steinberg, V. (2007) Longest relaxation times of double-stranded and single-stranded DNA. *Macromolecules*, **40**, 2172–2176.

Index

a
AC electro-osmosis (ACEO)   163
action–action–angle fluid flows   6
adiabatic diffusion   19, 20, 30
adiabatic invariant (AI)   6, 8, 9, 16, 27
– destruction   12
– diffusion, quantitative properties   12
– improved   21, 22
– jump (*See* jumps)
advection–diffusion equation   113
advection–diffusion system   116
advection-reaction-diffusion (ARD) systems   112
– behavior   120
– equation   118
– front propagation   125–127
– local behavior   120–122
– local pattern formation   122
– principles   118–120
advective mechanism   115
AI. *See* adiabatic invariant (AI)
angular rotation rate   36
angular speed   37
angular velocity   37, 42
approximate approach   104
Archimedean spiral   36
asymptotic flow
– analytical solution   45
– Poincaré map   46
autocatalytic reaction, pattern   121
autonomous flows
– 3D autonomous base flows   8
– flow structure   18
– frequency of unperturbed flow   17
– passages through resonances in   17–19
axis crossings   17
azimuthal velocity   36

b
Belousov-Zhabotinsky (BZ) reaction   112, 117, 118, 123
blinking vortex flow   120
Boger fluids   150
Boltzmann's constant   152
boundary layers
– development   104
Brownian motion   101
Brownian tracers   102

c
canonical flows
– viscoelastic instabilities in   156–160
capillary break up extensional rheometer (CaBER)   145
capture into resonance   20, 21
– captured motion   21
– phase portrait   20
– as probabilistic phenomenon   21
carboxy methyl cellulose (CMC)   142
Carreau–Yasuda model   148
Cartesian coordinates   36, 38, 39
Cauchy–Green strain tensor   66
chaotic domain   7, 10–12
– boundary of   15, 23
– size of   11
chaotic flow   71, 92, 97, 99, 102, 106, 107, 131, 141, 165
chemical reactions   111, 117, 118, 138, 146
Chilcott–Rallison approximation   150
connected networks models   123
conserved quantity   6
consistency index   148
continuity equation   146
continuously stirred tank reactor (CSTR)   117

Transport and Mixing in Laminar Flows: From Microfluidics to Oceanic Currents,
First Edition. Edited by Roman Grigoriev.
© 2012 Wiley-VCH Verlag GmbH & Co. KGaA. Published 2012 by Wiley-VCH Verlag GmbH & Co. KGaA.

continuum approximation 136, 137
convection–diffusion equation 104, 113
convective–diffusive process 103
coordinate system 141
corotating vortices 116
Couette cell 36, 144
Couette flow 141, 148
Coulomb forces 152
Criminale–Eriksen–Filbey (CEF) equation 148
cross-slot geometry 158, 159

d
Damkohler number 119
Dean flows 97
Deborah number 135, 136
Debye–Hückel parameter 153
deformation 46
– DNA molecule 164
– extensional 144
– of fluid elements 36, 37
– gradient 67, 73, 74, 76, 82
– grid 72
– oscillatory 144
– rates 136, 161
– spherical fluid element 69
– tensor 147, 148
diffusion coefficients 24, 138, 139, 152
diffusion equation 100
diffusion length 35, 36
diffusion-limited flux 93
dipole component 13
direct numerical simulations (DNS) 161
distribution function 24
DNA molecules 158, 164. See also deformation
– analysis 135
– relaxation process 165
domain boundaries 17
drift velocity 123

e
Eckman pumping 25, 26
eigenvalue equation 42
elasticity number 135
elastic turbulence 141, 160, 161
– transition to 160
electric double layer (EDL) formation 132
electrochemical methods, advantages 97
electrochemical reactions 95–97
electrochemical systems 91, 103
electrokinetic instabilities (EKI) 162
electro-osmosis (EO) 132, 162, 163
– equations 151–153

– principle 151
electro-osmotic flows (EOF) 132, 151
– complex fluids 163
– disadvantages 162
– polymer solutions 163
– velocities 147
electrophoresis 133, 162, 164, 165
energy exchange 7
EO. See electro-osmosis (EO)
EOF. See electro-osmotic flows (EOF)
Eulerian frame, numerical solutions 100, 101
extensional viscosity 144, 145
extra-stress tensor 149, 150

f
Faraday's constant 97
FENE-MCR model 149, 150
FENE-P constitutive equations 163
FENE-P model 151
Fick's law 100, 146
filament stretching extensional rheometer (FiSER) 145
finite element method (FEM) 100
finite perturbations 16
finite-size Lyapunov exponent (FSLE) 65
finite-strain (FS) field 64
– computation 64, 65
finite-time Lyapunov exponent (FTLE) 63, 65–67, 113, 120. See also Lagrangian coherent structures (LCSs)
– backward and forward time FTLE fields 65
– Cauchy–Green strain tensor 66
– computation 72–76, 80, 81
– Euclidean norm 66
– grid 78
– Lyapunov exponents 66
– ridges 67, 68
– – extraction 81
– – lock-in and enhance 79
FKPP approach 126
FKPP theory 117, 125
flip–flop microfluidic device 158
flow
– in microchannels 5
– in tilted rotating tank 39
– vorticity 36
fluid elasticity 154
fluid–fluid interfaces 91
fluid viscosities 12
Fokker–Planck equation 11
forcing methods 162–165
Fourier coefficients 29
Fourier's law of heat conduction 154

Index

free-surface flows   5
FTLE. *See* finite-time Lyapunov exponent (FTLE)

g

generalized Newtonian fluids (GNF)   147, 148
generalized viscosity function, values of parameters   148
Giesekus model   149, 151
Gordon–Schowalter derivative   150, 151
governing equations   146–154
– for flows of complex fluids   141
– for Newtonian and non-Newtonian fluid flow   131
– for nonisothermal flows   153
Grætz analysis, of interfacial transfer   103
Grætz prediction   106

h

Hamiltonian function   9
Hamiltonian systems   7, 9, 10, 12
heat flux   154
heteroclinic tangle   62
high-molecular-weight polymeric solutions   137
high-order resonances   29, 30
homogeneous fluid   36
– KAM region the mode of transport in   36
– laminar flow experiments   48, 51–53
– low Reynolds number experiments   47
hyperbolicity time (HT)   65

i

improved AI *vs.* original AI,   22
induced-charge electro-osmosis (ICEO)   163
inelastic non-Newtonian fluids. *See* generalized Newtonian fluids (GNF)
inhomogeneous fluids
– laminar flow experiments   48, 53, 54
integrable system   6
interfacial mass transfer measurements   95
interfacial transfer processes   100
– modeling approaches   100–107
interfacial transport   91
interphase mass transfer   98
– from droplets in microfluidic flows   99
– renewal theory   104
invariant tori   6, 23

j

Johnson-Segalman (JS) model   149
Josephson junctions   7
jumps
– accumulation   7, 10, 11
– AI associated   7, 16
– at boundaries of system   17
– decay   29
– distribution   11, 17
– due to scattering   22
– and ergodicity, correlations   12
– between first- and second-layer boundaries   22, 23
– magnitude   31
– statistical properties   7, 12
– types   17

k

KAM regions   35
KAM surface   35, 38, 56
KAM tori   6
Knudsen number   137

l

lab-on-chip devices   165
Lagrangian coherent structures (LCSs)   59, 114
– applications   83
– computation   72–75, 82, 83
– – grid-based   74, 75
– – integration time   75–79
– detection   75
– – HT approach   70
– extraction   79–81
– hyperbolic   77
– identification   72, 73
– as most repelling/attracting material surfaces   64
– robustness   82
– spurious   78
– *vs.* stable/unstable manifolds   63
– *vs.* backward time FTLE   71
Lagrangian frame, numerical solutions   101–103
laminar flows   5, 44
– heat transfer measurement   97
– homogeneous fluid   51–53
– interfacial transfer processes   100
– microfabrication using UV laminar flow patterning   133
– Newtonian fluids in ducts   138
laminar mixing effects
– advection–diffusion equation   113
– ARD systems
– – front propagation in   125–127
– – local behavior   120–122
– background   113–118
– impurities, long-range transport   115–117

– oscillating reactions, synchronization 122–124
– reaction–diffusion systems 117, 118
– on reaction fronts and patterns 111–113
– short-range mixing 113–115
large amplitude oscillatory shear flow (LAOS) 145
LCSs. *See* Lagrangian coherent structures (LCSs)
Lévy flights 116, 124
– trajectories 122
life-cycle processes 111
linear velocity distributions 36
liquid–liquid interface 91
local mass transfer coefficient 93
long-time dynamics 19, 20
Lorentz force 147
low-order resonance 30
Lyapunov exponents 12, 37, 66

m

macroscopic coiled pipes
– heat transfer from 98
mass transfer rates in microfluidic potential cells
– electrochemical measurements 96
metrics of mixing
– rate of mixing D 11
– volume of the chaotic domain, V_c 10
microfabrication techniques, recent developments 134
microfluidic device
– *flip–flop* microfluidic device 158
– junction of two Newtonian fluid streams in 138
– length scales 139
– Newtonian fluid streams, junction 138
– series of connected triangular elements 157
microfluidic flows 131–166
– complex fluids in 135, 136
– Lagrangian method, interfacial transfer from 92
microfluidics 5, 131–137
– applications 133–135
– basic principles 131–135
– characteristics 133
– Reynolds number 133
microfluidics mixing 138–141
– challenges 138, 139
– enhancement, methods 139–141
microfluidic T-junction geometry 158
microfluidic T-shaped channel 164
mixing domain 11, 12, 23, 29–31

mixing processes 99
mode-locking 125, 126
Moffatt vortices 44
momentum equations 146, 147
momentum transport 35

n

Navier–Stokes equations 25, 100
near-integrable 5
near-integrable flows, general properties 8
near-integrable systems 5
Nernst–Planck equations 152, 153
Newtonian fluids 38, 131, 137, 138, 144, 145
– flow in complex geometries 162
– generalized model 148
– micromixing in 140
Newtonian plateau 142
nonautonomous flows
– averaging 26, 27
– – evolution equations 26
– – equations 24, 25
– passages through resonances 24, 25
– two-dimensional 26
– unperturbed flow 25, 26
nonautonomous systems 9
nondimensional mass transfer coefficient 94
nonlinear equations of motion 42, 45
non-Newtonian fluid 56, 131, 141, 142
– EO flows of 163
– as incompressible fluids 148
– inelastic 147
– rheology 141
non-Newtonian viscoelastic fluids 141–145, 153
– normal stresses 143
– shear viscosity 142, 143
– storage and loss moduli 144
numerical simulations 16, 24
numerical solutions
– in Eulerian frame 100, 101
– evolution of distribution of concentration 93, 94
– Poincaré map 46
Nusselt number 94, 97

o

Oldroyd-B constitutive equation 161
Oldroyd-B models 150, 161
oscillating BZ reaction
– sequences of images 124
oscillating/drifting vortex chain 123
oscillating vortex chain flow 124–126
Ostwald de Waele power law 148

p

partial averaging 28
particle image velocimetry (PIV) 65, 71
passive mixing 154–161
Péclet numbers 91, 113, 119, 138
periodic shearing 36
perturbation theory 5, 6, 13
perturbed-phase trajectory 14
Peterlin's approximation 151
Phan-Thien–Tanner model (PTT) 149
phase trajectory 14
phase waves 118
planar Cartesian coordinate 36
plane Couette flow 141
Poincaré mapping 38, 46, 60
Poincaré sections 12
Poiseuille flow 94, 105
Poisson–Boltzmann–Debye–Hückel model 153
Poisson–Boltzmann model 153
polar geometry 37
polyacrylamide (PAA) 143, 151, 163
polyethylene oxide (PEO) polymer 157
power index 148
Prandtl number 91
pressure-driven flows 132
– parabolic velocity profile 132
probability density function (PDF) 11
protein binding 93, 95, 100

q

quadrupole flow 13

r

radial velocity 36
rate of deformation tensor 147
reaction–diffusion patterns 119
reaction-diffusion (RD) systems 112, 117, 118
Reiner–Rivlin equation 149
relative dispersion (RD) 65
resonance condition 27
resonance phases 12
resonance surfaces 7–9, 27
resonant phenomena 27–29
Reynolds number 38, 44, 91, 97, 112, 127, 133, 135, 136, 140, 141, 165
– inhomogeneous flow 38
Reynolds stresses 161
rheological constitutive equation 145, 147–151, 165
– generalized Newtonian fluid model 148
– viscoelastic stress models 148–151
rotating tank motion 37
rotational vortex 37
rotation rates 35, 36, 38
Runge–Kutta scheme 101
Runge–Kutta algorithm 45

s

scattering on resonance 19, 20
– jump, magnitude of 19
– phase portrait 20
Schmidt number 91
Sentmanat device 145
separatrix crossings 12, 17
– component flows 13, 14
– in volume-preserving systems 12–14
separatrix surface 9
– dynamics 15, 16
– vs. resonance surface 9
serpentine channels 98
shear viscosity 142, 143
Sherwood number 94, 96, 97, 101
2D singular surfaces 9
small amplitude oscillatory shear (SAOS) flow 143, 144, 150
solid-body rotation 37, 38, 43
solid–liquid interfaces 91
spatial resolution 95
standard electrokinetic model 147
stirred laminar flows
– electrochemical reactions 95–97
– experimental methods 95–99
– experimental observations, summary 99
– heat transfer from macroscopic coiled pipe 96, 97
– interfacial transfer from 91
– interphase mass transfer from droplets 97–99
– macrotransport approach 103
– modeling approaches 100–107
– phenomena and definitions 91–95
– protein binding 95
– theoretical approaches 103–106
Stokes flow equations 38
stress coefficient function 150
stress components 40
stress difference coefficient 143
stretching
– DNA molecules 164
– fluid elements 114
– impurities in flow 120
– long molecules 158
– periodic shear 36
– polymer molecules 161
– rates 78, 79

– for stirring of fluid element 36
– in tilted rotating tank 45
– use of stretching fields 120
superdiffusive transport 116
surface plasmon resonance (SPR) 95
surface tension 5, 12, 13, 38, 132, 165
– confining effect 5
– gradients 166
surface-to-volume ratio 132

t

Taylor–Couette flow 155
Taylor flow 13
temperature coefficients 12, 13
temporal inhomogeneity in shear 36
thermal energy equation 153, 154
thermocapillary effect 13
tilt angle 40, 43, 49, 51, 53–55
tilted-rotating tank
– equations of motion, asymptotic limit 40
– general solution of reduced equation 41
– liquid rotation axis in asymptotic limit 41
– model equation 38–40
time-dependent perturbation 26, 27
time-varying processes 122
T-junction arrangements 164
transfer coefficients 107
trigger waves 118
Trouton ratio 145
turbulence 112
turnstile mechanism 116

u

UCM fluids, creeping flow in smooth contractions 159
unperturbed flow 5, 14, 25, 26
– frequency 17
unperturbed system 6, 8, 9, 14, 18, 26
– variables 18
upper-convected Maxwell (UCM) model 149, 150, 158
UV laminar flow patterning 133

v

velocity field 8, 59, 60, 66, 72, 73, 78, 82, 113, 154
viscoelastic constitutive equations
– model parameters 149

viscoelastic fluids
– canonical flows, viscoelastic instabilities 156–160
– continuity and momentum equations 146, 147
– continuum approximation 136, 137
– elastic turbulence 160, 161
– electro-osmosis 162, 163
– – equations 151–153
– electrophoresis 164, 165
– extensional viscosity 144, 145
– forcing methods 162–165
– governing equations 146–154
– microfluidic flows 131–137, 131–166
– – complex fluids in 135, 136
– – mixing in 138–141
– non-Newtonian viscoelastic fluids 141–145
– objectives and organization 131
– passive mixing 154–161
– rheological constitutive equation 147–151
– rheological properties 145
– small amplitude oscillatory shear (SAOS) flow 144
– thermal energy equation 153, 154
– underlying physics 155
viscosity coefficient 147
visual cortex, spreading depression 118
3D volume-preserving autonomous systems 7
volume-preserving kinematic model 17
volume-preserving systems 10, 12
volume-to-surface ratio 133
vortex dipole 61, 62
vorticity
– associated with flow rotation 36
– spatially inhomogeneous 37
– vertical 40

w

Watts/Strogatz small-world network model 124
Weissenberg effect 143
Weissenberg number 135, 136
Williams–Landel–Ferry (WLF) equation 149

z

zeta potential 147